# Design and Performance of Embankments on Very Soft Soils

# Design and Performance of Embankments on Very Soft Soils

## Márcio de Souza S. Almeida
*Graduate School of Engineering, Federal University of Rio de Janeiro, Rio de Janeiro, Brazil*

## Maria Esther Soares Marques
*Department of Fortification and Construction Engineering, Military Institute of Engineering, Rio de Janeiro, Brazil*

CRC Press
Taylor & Francis Group
Boca Raton  London  New York

CRC Press is an imprint of the
Taylor & Francis Group, an **informa** business

A BALKEMA BOOK

CRC Press
Taylor & Francis Group
6000 Broken Sound Parkway NW, Suite 300
Boca Raton, FL 33487-2742

First issued in paperback 2018

ISBN 13: 978-0-415-65779-2 (hbk)
ISBN 13: 978-1-138-07693-8 (pbk)

Published by:   CRC Press/Balkema
                P.O. Box 11320, 2301 EH, Leiden, The Netherlands
                e-mail: Pub.NL@taylorandfrancis.com
                www.crcpress.com – www.taylorandfrancis.com

Visit the Taylor & Francis Web site at
http://www.taylorandfrancis.com

and the CRC Press Web site at
http://www.crcpress.com

To Maria, Adriana and Leandro for their continued
support over the years.
Márcio

To my family and students.
Esther

# Table of contents

# Preface

Even if it is an important topic in geotechnical engineering, embankments on soft or very soft soils have been the subject of few books and, to my knowledge, none recently published. This book "Design and Performance of embankments on Very Soft Soils" is thus very welcome.

The authors, Márcio Almeida and Esther Marques, have a long experience with soft soils and embankments. Indeed both did their Ph.D. on related topics. They also have an excellent knowledge of advanced soil mechanics and of new technologies for both characterizing soft soil deposits and solving settlement or stability problems, as well as field monitoring and interpretation. The book reflects this state-of-the-art knowledge. Soils are described using modern concepts of yielding and yield curves; sampling quality is considered; the use and interpretation of DMT, T-bar and piezocone soundings are described. Technologies for reducing and/or accelerating settlements and for improving stability are also described. In particular, emphasis is put on "embankments on pile-like elements" and on "vacuum preloading" with which the authors have very good experience.

With this book in English, in addition to the general technical aspects previously mentioned, Professors Márcio Almeida and Esther Marques offer the geotechnical community the remarkable and unique Brazilian experience with embankments on very soft organic soils. Very nice contribution!

Serge Leroueil,
July 2013

# About the authors

Márcio Almeida earned his Civil Engineering degree at the Federal University of Rio de Janeiro, in 1974 and obtained his MSc at COPPE/UFRJ in 1977 when he joined COPPE as Assistant Lecturer. Marcio got his PhD from the University of Cambridge, UK in 1984. Then he returned to UFRJ and in 1994 became Professor of Geotechnical Engineering. His postdoc was at Italy (ISMES) and NGI, Norway in the early 1990s and he was also visiting researcher at the universities of Oxford, Western Australia and ETH, Zurich. He is currently one of the leading researchers of the National Institute of Science and Technology – Rehabilitation of Slopes and Plains (INCT-REAGEO). He has been the Director of COPPE's MBA "Post-Graduate Program in Environment" since 1998. He has published numerous articles in journals and conferences in Brazil and abroad and has supervised over 60 doctoral and master dissertations. He received the Terzaghi and Jose Machado awards from the Brazilian Association of Soil Mechanics and Geotechnical Engineering (ABMS). His experience ranges from soft clay engineering, environmental and marine geotechnics, site investigation, physical and numerical modeling as well as extensive experience in geotechnical consulting.

Esther Marques holds a degree in Civil Engineering – emphasis in Soil Mechanics, from Federal University of Rio de Janeiro. She obtained her MA and PhD in Civil Engineering from COPPE/UFRJ, with researches conducted at Université Laval, Canada. She worked at Tecnosolo and Serla and was a researcher at COPPE/UFRJ from 2001 to 2007. She is currently an associated professor at the Military Institute of Engineering, where she teaches undergraduate and graduate Transportation Engineering and Defence Engineering. She has experience in Civil Engineering with emphasis in Soil Mechanics, working mainly with the following: laboratory testing, field-testing, instrumentation, soft soils behavior, embankments on soft soils and environmental geotechnics.

# Acknowledgements

I owe my geotechnical background to Fernando Barata, Costa Nunes, Dirceu Velloso, Márcio Miranda, Jacques de Medina and Willy Lacerda, among several others from a great host of professors at UFRJ. I learned Critical States Soil Mechanics and Centrifuge Modeling during my PhD at Cambridge University, with Andrew Schofield, Dick Parry, David Wood, Malcolm Bolton and Mark Randolph. Mike Gunn and Arul Britto gave important support in those early years of Cam-clay Numerical Modeling with CRISP. In subsequent years, Peter Wroth, Gilliane Sills and Chrisanthy Savvidou were remarkable in scientific collaborations with Oxford and Cambridge. Mike Jamiolkowski and Tom Lunne were very receptive during the postdoctoral sabbatical in Italy (ISMES) and Norway (NGI), respectively, and Mark Randolph and Martin Fahey years later in Australia (UWA).

I also thank the many colleagues, who were important for the exchange of experiences and collaborations during all these years: Antonio Viana da Fonseca, Ennio Palmeira, Fernando Danziger, Fernando Schnaid, Flávio Montez, Francisco Lopes, Ian Martin, Jarbas Milititsky, Jacques Medina, Leandro Costa Filho, Luc Thorel, Luiz Guilherme de Mello, Maria Cascão, Maria Claudia, Maurício Ehrlich, Mike Davies, Osamu Kusakabe, Roberto Coutinho, Sandro Sandroni, Serge Leroueil and Sarah Springman, among many others.

Finally, I thank my research students for imparting knowledge during their master's and doctoral research, among whom I highlight Esther Marques, Henrique Magrani, José Renato Oliveira and Marcos Futai, for the continued collaboration, and Mario Riccio, for his support in proofreading parts of this book.

Márcio Almeida

I owe the awakening of my interest in Geotechnics to the professors at the Polytechnic School of UFRJ: Maurício Ehrlich, Fernando Barata and Willy Lacerda, among others.

I thank the colleagues from Tecnosolo for the opportunity to share with them the experiences of geotechnical engineering practices at the beginning of my career, under the baton of Prof. Costa Nunes.

When continuing my graduate studies at COPPE-UFRJ, socializing with professors really encouraged me to stay in academia. I thank the teachers Márcio Almeida and Ian Schumann for the guidance and friendship during this time. I had the opportunity to develop research under the guidance of Serge Leroueil, to whom I am thankful for the welcome at Laval University.

I thank colleagues from COPPE-UFRJ, especially Prof. Márcio Almeida, for the opportunity to work on research projects that have contributed to my academic enrichment.

I thank colleagues from IME for the friendship and support in the courses and works, particularly Professor Eduardo Thomaz and colleagues of SE-2. To my students, in addition to the dedication, I thank them for the challenge, which is the motivation for improving daily.

Esther Marques

# List of symbols

## GEOMETRIC PARAMETERS

| | |
|---|---|
| $A$ | area of draining mattress cross section, referring to a line of drains (Chapter 5) – $m^2$ |
| $A$ | area of unit cell (Chapter 7) – $m^2$ |
| $a$ | largest dimension of a rectangular PVD (Chapter 5) – m |
| $A_c$ | granular column area – $m^2$ |
| $a_c$ | normalized granular column area or replacement ratio of granular columns – $m^2$ |
| $A_n$ and $A_t$ | areas of the cone tip (Chapter 2) – $m^2$ |
| $A_s$ | area of soil (clay) in the unit cell of the granular column – $m^2$ |
| $a_s$ | normalized area of the soil around the granular column in the unit cell – $m^2$ |
| $B$ | average width of the embankment platform (Chapter 6) – m |
| $b$ | smallest dimension of a rectangular PVD (Chapter 4) – m |
| $b$ | width of the embankment platform (Chapter 4) – m |
| $b$ | width of the pil cap (Chapter 7) – m |
| $d$ | diameter of granular column – m |
| $D$ | thickness of the clay layer (Chapter 6) – m |
| $D$ | vane diameter (Chapter 2) – m |
| $D_{50}$ and $D_{85}$ | particle diameter for which 50% and 85% of soil mass is finer, respectively – m |
| $d_e$ | diameter of influence of a drain or equivalent diameter of a granular column considering an unit cell – m |
| $d_e$ | external diameter of the piezocone probe (Chapter 2) – m |
| $d_i$ | inner diameter of the piezocone probe (Chapter 2) – m |
| $d_m$ | equivalent diameter of the driving mandrel – m |
| $d_m^*$ | equivalent diameter of the footing mandrel – m |
| $d_s$ | diameter of the area affected by disturbance – m |
| $d_w$ | cylindrical-shaped drain diameter or equivalent diameter of a PVD with rectangular section – m |
| $H$ | vane height – m |
| $h_{adm}$ | allowable embankment height adopted in the design – m |
| $h_{blanket}$ | thickness of the drainage blanket – m |
| $h_c$ | height of granular column – m |

$h_{cd}$    height of head loss in the drainage blanket – m

$h_{clay}$    thickness of the clay layer – m

$h_{cr}$    critical height or height of the collapse of non-reinforced embankment – m

$h_d$    draining distance – m

$h_{emb}$    thickness or height of the embankment – m

$h_f$    final fill thickness, including surcharge – m

$h_{fs}$    total fill thickness, including surcharge – m

$h_s$    thickness of embankment surcharge – m

L    characteristic length of the vertical drain (Chapter 5) – m

l    distance between drains or granular columns (Chapters 5 and 7) – m

L    distance between inclinometer measurements (Chapter 8) – m

L    horizontal length of failure surface (Chapter 6) – m

l    thickness of a rectangular mandrel – m

$L_{anc}$    anchoring length of reinforcement – m

n    slope inclination

$O_{50}$    particle diameter for which 50% of the soil passes through the geotextile – m

$O_{90}$    geotextile filter opening, defined as the largest grain of soil able to pass through it – m

R    piezocone radius – m

r    the radial distance measured from the drainage center to the point considered – m

$r_c$    initial radius of granular column – m

$r_e$    unit cell radius – m

$r_{geo}$    radius of the geosynthetic cylinder – m

$r_w$    cylindrical drain radius or equivalent radius of rectangular PVD – m

s    distance between axis of pile or columns in piled embankments – m

$V_h$    estimated volume of soil mass dislocated calculated from measured horizontal displacements – $m^3$

$V_v$    estimated volume of soil mass dislocated calculated from measured settlements – $m^3$

w    width of a rectangular mandrel – m

$X_T$    distance between the foot of the slope and the point where the circle intercepts the reinforcement – m

z    depth of the analyzed soil regarding the level of the natural terrain (Chapter 4 and 5) – m

z    depth of the inclinometer reading (Chapter 8) – m

$z_{clay}$    depth of the rupture surface within the clay layer (wedge method) – m

$z_{clay}$    depth of the sample (Chapter 2) – m

$z_{crack}$    depth at which the crack develops in the embankment – m

$\Delta r_c$    variation of the column radius – m

$\Delta r_{geo}$    variation of the geosynthetic radius – m

$\alpha$    area ratio of cone tip ($= A_n/A_t$)

$\theta$    inclination angle of the inclinometer tube (Chapter 8) – °

$\theta$    rotation angle measured by the vane test (Chapter 3) – °

$\theta_{max}$    rotation angle measured by vane test regarding maximum torque (Chapter 3) – °

$\Lambda$    dimensionless critical state parameters ($= 1 - C_s/C_c$)
$\phi$    diameter of the Shelby tube (Chapter 2) – m
$\phi_{sample}$    diameter of the sample – m

## MATERIAL PARAMETERS

$a_v$    soil compressibility index – $m^2/kN$
$B_q$    piezocone parameter for soil classification
$c$    cohesion – $kN/m^2$
$c'$    effective cohesion – $kN/m^2$
$c'_c$    effective cohesion of the column of the granular material – $kN/m^2$
$c'_s$    cohesion of the soil around granular column for drained condition – $kN/m^2$
$C_c$    compression index
$c_d$    mobilized cohesion fill – $kN/m^2$
$c_{emb}$    cohesion of the fill – $kN/m^2$
$c_h$    coefficient of consolidation for horizontal draining (flow) – $m^2/s$
$c_m$    weighted cohesion of soil/granular column system – $kN/m^2$
CR    compression ratio
$C_R$    recompression index
$c_s$    cohesion of the soil around granular column – $kN/m^2$
$C_s$    swelling or recompression (or unloading-reloading) index
$c_v$    coefficient of consolidation for vertical draining (flow) – $m^2/s$
$c_{vfield}$    coefficient of vertical consolidation calculated from monitoring data – $m^2/s$
$c_{vlab}$    coefficient of vertical consolidation obtained from laboratory tests – $m^2/s$
$c_{vpiez}$    computed coefficient of vertical consolidation from piezocone dissipation test, corrected for flow direction – $m^2/s$
$C_\alpha$    coefficient of secondary compression
$E$    modulus of elasticity or Young modulus – $kN/m^2$
$E^*$    modulus of elasticity or Young's modulus of granular encased column (Chapter 7) – $kN/m^2$
$E'$    modulus of elasticity or Young's modulus (Chapter 7) – $kN/m^2$
$E_c$    modulus of elasticity of granular column – $kN/m^2$
$e_o$    initial void ratio of sample in a laboratory
$E_{oed}$    oedometer modulus (or confined module) – $kN/m^2$
$E_{oeds}$    oedometer modulus of the soil for a given stress – $kN/m^2$
$E_{oedsref}$    reference oedometer modulus of the soil (obtained for stress $P_{ref}$) – $kN/m^2$
$E_s$    modulus of elasticity of the soil around the granular column – $kN/m^2$
$E_u$    modulus of elasticity (Young's modulus) for the undrained condition – $kN/m^2$
$E_{u50}$    secant $E_u$ modulus for 50% stress of maximum stress deviation – $kN/m^2$
$e_{vo}$    void ratio corresponding to the *in situ* initial effective vertical stress
$G_{50}$    shear stress modulus for small deformations for 50% of maximum shear stress – $kN/m^2$
$G_o$    shear stress modulus for small deformations range (or $G_{max}$) – $kN/m^2$
$G_s$    density of the grains – $kN/m^2$
$I_P$    plasticity index

| | |
|---|---|
| $I_R$ | soil rigidity index $(= G/S_u)$ |
| $J$ | stiffness modulus of geosynthetic or reinforcement – kN/m |
| $J_R$ | nominal stiffness modulus of geosynthetic or reinforcement – kN/m |
| $k$ | coefficient of permeability – m/s |
| $K'$ | bulk modulus |
| $k'_h$ | coefficient of horizontal permeability of the area affected by disturbance – m/s |
| $K_{ac}$ | coefficient of active earth pressure of granular column |
| $K_{aclay}$ | coefficient of active earth pressure of clay |
| $K_{aemb}$ | coefficient of active earth pressure $K_{pclay}$ coefficient of passive earth pressure of clay |
| $k_{blanket}$ | coefficient of permeability of the material of the drainage blanket-m/s |
| $k_h, k_v$ | coefficient of horizontal and vertical permeability respectively – m/s |
| $k_{h0}, k_{v0}$ | coefficient of horizontal and vertical permeability at *in situ* stress respectively – m/s |
| $K_o$ | coefficient of earth pressure at rest |
| $K_{oL}$ | value of $K_o$ at the limit of zones at indifferent equilibrium and with secondary compression |
| $K_{os}$ | coefficient of earth pressure at rest $(= 1 - \sin \phi')$ in excavation method |
| $K_{os}^*$ | increased $K_o$ in displacement method |
| $K_{pemb}$ | coefficient of passive earth pressure of the fill material $m_v$ coefficient (or volumetric variation) of vertical compressibility |
| $M$ | inclination of the Critical State Line in the $p'$-$q$ plot |
| $m_v$ | coefficient of volume compressibility – m$^2$/kN |
| $S_{clay}$ | mobilized shear force of soft clay at a given plane – kN/m$^2$ |
| $S_t$ | clay sensitivity |
| $S_u$ | undrained strength of clay – kN/m$^2$ |
| $S_{uh}$ | undrained strength of clay in horizontal direction (vane test) – kN/m$^2$ |
| $S_{uo}$ | undrained strength of clay at soil/embankment interface – kN/m$^2$ |
| $S_{ur}$ | undrained remolded strength of clay – kN/m$^2$ |
| $S_{uv}$ | undrained strength of clay in vertical direction (vane test) – kN/m$^2$ |
| $V$ | specific volume |
| $w_L$ | liquidity limit |
| $w_n$ | natural water content *in situ* |
| $w_P$ | plasticity limit |
| $\Delta e_{vo}$ | variation of void ratio from the start of the test to the effective vertical stress *in situ* |
| $\gamma'_c$ | specific submerged weight of granular material of the column – kN/m$^3$ |
| $\gamma'_{emb}$ | specific submerged (effective) weight of embankment – kN/m$^3$ |
| $\gamma'_s$ | specific submerged weight of soil around the column – kN/m$^3$ |
| $\gamma_c$ | specific weight of granular column material – kN/m$^3$ |
| $\gamma_{clay}$ | specific weight of clay – kN/m$^3$ |
| $\gamma_{emb}$ | specific weight of embankment – kN/m$^3$ |
| $\gamma_m$ | average specific weight of soil/granular column system – kN/m$^3$ |
| $\gamma_{nat}$ | specific weight of natural soil – kN/m$^3$ |
| $\gamma_s$ | specific weight of soil around the granular column – kN/m$^3$ |

$\gamma_w$     specific weight of water – $kN/m^3$
$\mu$     viscosity
$\Phi$     internal friction angle of the soil – °
$\Phi'$     effective internal friction angle of the soil – °
$\Phi'_{cs}$     internal friction angle of the soil at Critical State- °
$\Phi_c$     internal friction angle of the granular material of the column – °
$\Phi_d$     mobilized friction angle of the fill material – °
$\Phi_{emb}$     internal friction angle of fill material – °
$\Phi_m$     weighted internal friction angle of the soil/granular column system – °
$\Phi_s$     internal friction angle of the soil around the granular column – °
$\kappa$     swelling index of isotropic consolidation in the e vs ln p' plot
$\lambda$     compression index of isotropic consolidation in the e vs ln p' plot
$\nu$     Poisson's ratio
$\nu_u$     Poisson's ratio for undrained conditions
$\nu'$     Poisson's ratio in terms of effective stress
$\nu_s$     Poisson's ratio of the soil


## DISPLACEMENTS, FORCES, PRESSURES, STRAINS, STRESSES AND VELOCITIES

d     distortion along the inclinometer tube
$F_s$     lateral resistance measured at piezocone normalized by the net tip resistance – $kN/m^2$
$f_r$     lateral friction (piezocone test) – $kN/m^2$
$F_r$     tensile strength of geosynthetic – $kN/m$
$f_s$     lateral resistance of the cone – $kN/m^2$
$P^*$     active earth pressure – $kN/m^2$
$p'$     mean effective stress – $kN/m^2$
$p'_c$     preconsolidation or yield stress at the isotropic consolidation line – $kN/m^2$
$p'_f$     mean effective stress at failure – $kN/m^2$
$P_{aclay}$     active pressure in the soft clay layer – $kN/m^2$
$P_{aemb}$     active pressure on fill layer – $kN/m^2$
$P_{pclay}$     passive pressures on soft clay layer – $kN/m^2$
$P_{pemb}$     passive pressures in fill layer – $kN/m^2$
$P_{ref}$     reference stress (Chapter 7) – $kN/m^2$
$P_{ref}$     shear force at the base of embankment (Chapter 6) – $kN/m^2$
q     deviator stress or shear stress – $kN/m^2$
q     surcharge – $kN/m^2$
$q_b$     tip resistance measured in cylindrical bar test (T-bar) – $kN/m^2$
$q_c$     tip resistance measured in cone test – $kN/m^2$
$q_f$     deviatory stress or shear stress at failure – $kN/m^2$
$q_t$     corrected tip resistance of piezocone test – $kN/m^2$
$Q_t$     net tip resistance (piezocone test) normalized by the total stress
r     settlement rate – m/s
s(t)     settlements over time – m

| | |
|---|---|
| $s_\infty$ | settlement at infinity – m |
| $s_i$, $s_{i+1}$ | settlements at time $t_1$ and $t_{1+1}$, respectively – m |
| T | tensile force at reinforcement (Chapter 6) – kN/m |
| T | tension in the geogrid – kN/m |
| T | torque measured in vane test (Chapter 3) – kN·m |
| $T_{anc}$ | anchoring strength of the reinforcement – kN/m |
| $T_{lim}$ | limit tensile force at the reinforcement – kN/m |
| $T_{max}$ | maximum torque measured in vane test – kN.m |
| $T_{mob}$ | mobilized tensile force at the reinforcement – kN/m |
| $T_r$ | nominal tensile strength of the reinforcement (geosynthetic) – kN/m |
| u | pore pressure – kPa |
| $u_0$ | initial hydrostatic pore pressure at a given depth – kPa |
| $u_1$ | pore pressure measured on the face of the cone at the given depth – kPa |
| $u_2$ | pore pressure measured on the base of the cone at a given depth – kPa |
| $u_{50\%}$ | pore pressure corresponding to the consolidation percentage equal to 50% at a given depth – kPa |
| $u_i$ | pore pressure at the start of dissipation test at a given depth – kPa |
| $\Delta$ | d variation of distortion measured in inclinometer tube |
| $\Delta F_R$ | increase in geosynthetic strength of encased granular column – kN/m |
| $\Delta h$ | final primary consolidation settlements (infinity) – m |
| $\Delta h(t)$ | primary settlement for a given time t – m |
| $\Delta h_a$ | primary consolidation settlements – m |
| $\Delta h_{adp}$ | virgin primary consolidation settlements – m |
| $\Delta h_{arec}$ | primary recompression settlements – m |
| $\Delta h_c$ | granular column settlement (Chapter 7) – m |
| $\Delta h_f$ | primary settlement due to increased vertical stress $\Delta\sigma_{vf}$ – m |
| $\Delta h_{fs}$ | primary settlement due to increased vertical stress $\Delta\sigma_{vfs}$ – m |
| $\Delta h_i$ | immediate settlement (also called undrained or elastic settlement) – m |
| $\Delta h_{if}$ | settlements of working platform – m |
| $\Delta h_{max}$ | maximum settlements in the center line of the embankment – m |
| $\Delta h_s$ | enhanced or treated soil settlement (Chapter 7) – m |
| $\Delta h_{sec}$ | secondary compression settlements – m |
| $\Delta h_t$ | settlement on top of the piled embankment – m |
| $\Delta p'$ | mean effective stress variation – kN/m$^2$ |
| $\Delta q$ | deviatory stress or shear stress – kN/m$^2$ |
| $\Delta r_c$ | variation of column granular radius -m |
| $\Delta u$ | pore pressure variation – kPa |
| $\Delta u_{50}$ | pore pressure variation up to 50% of dissipation – kPa |
| $\Delta\sigma$ | total vertical stress increase – kN/m$^2$ |
| $\Delta\sigma_0$ | increase of vertical stress (embankment over columns) – kN/m$^2$ |
| $\Delta\sigma_{hc}$ | variation of horizontal stress acting on the granular column – kN/m$^2$ |
| $\Delta\sigma_{hdif}$ | horizontal stress difference (between column and soil with geosynthetic) – kN/m$^2$ |
| $\Delta\sigma_{hgeo}$ | horizontal stress variation on geosynthetic – kN/m$^2$ |
| $\Delta\sigma_{hs}$ | variation of horizontal stress acting on the soil around the granular column – kN/m$^2$ |

| | |
|---|---|
| $\Delta\sigma_v$ | increase in vertical stress – $kN/m^2$ |
| $\Delta\sigma_{vc}$ | increase of vertical stress on granular column – $kN/m^2$ |
| $\Delta\sigma_{vf}$ | applied vertical stress (for a specific embankment height) – $kN/m^2$ |
| $\Delta\sigma_{vfs}$ | increase of vertical stress due to fill thickness $h_{fs}$ – $kN/m^2$ |
| $\Delta\sigma_{vs}$ | increase of vertical stress in the soil around the granular column – $kN/m^2$ |
| $\delta_h$ | horizontal displacement – m |
| $\delta_{hmax}$ | maximum horizontal displacement – m |
| $\varepsilon$ | strain |
| $\varepsilon_a$ | admissible axial strain on reinforcement |
| $\varepsilon_r$ | radial strain |
| $\varepsilon_v$ | vertical strain |
| $\sigma$ | total stress – $kN/m^2$ |
| $\sigma'_{vo}$ | initial effective vertical stress *in situ* – $kN/m^2$ |
| $\sigma_1^*$ | stress before loading- $kN/m^2$ |
| $\sigma'_1, \sigma'_2, \sigma'_3$ | effective principal stress, major, intermediate and minor respectively – $kN/m^2$ |
| $\sigma_2^*$ | stress after loading – $kN/m^2$ |
| $\sigma'_a$ | effective axial stress – $kN/m^2$ |
| $\sigma'_h$ | effective radial or horizontal stress – $kN/m^2$ |
| $\sigma'_{ho}$ | initial effective horizontal stress *in situ* – $kN/m^2$ |
| $\sigma'_v$ | effective vertical stress – $kN/m^2$ |
| $\sigma_v$ | total vertical stress – $kN/m^2$ |
| $\sigma_v$ | vertical stress acting on the geosynthetic (Chapter 7) – $kN/m^2$ |
| $\sigma_{vaverage}$ | average vertical stress *in situ* from the instrumentation data – $kN/m^2$ |
| $\sigma'_{vc}$ | effective consolidation pressure of triaxial tests – $kN/m^2$ |
| $\sigma'_{vf}$ | final effective vertical stress – $kN/m^2$ |
| $\sigma'_{vm}$ | overconsolidation stress – $kN/m^2$ |
| $\sigma_{vo}$ | initial total vertical stress *in situ* – $kN/m^2$ |
| $\sigma_{voc}$ | initial vertical stress (without surcharge) of the column soil at a given depth – $kN/m^2$ |
| $\sigma_{vos}$ | initial vertical stress (without surcharge) of the soil around the column at a given depth – $kN/m^2$ |
| $\tau$ | shear stress at the base of embankment – $kN/m^2$ |
| $\nu_d$ | distortion rate |

## OTHER SYMBOLS

| | |
|---|---|
| $C_i$ | geosynthetic/soil interaction coefficient |
| DR | maximum settlement and maximum horizontal displacement ratio |
| F(n) | geometric factor in radial drainage, function of drains density |
| F | parameter of theTaylor and Merchant theory |
| $F_q$ | increase of F(n) value due to the hydraulic resistance of the drain in radial drainage |
| $FR_{DB}$ | partial reduction factor of T due to biological degradation |
| $FR_{DQ}$ | partial reduction factor of T due to chemical degradation |

| | |
|---|---|
| $FR_F$ | partial reduction factor of T due to creep in geosynthetic |
| $FR_I$ | partial reduction factor of T due to mechanical damage during installation |
| $F_s$ | increase in the value of $F(n)$ due to disturbance around the drain in radial drainage (Chapter 5) |
| $F_s$ | Safety Factor |
| i | hydraulic gradient |
| I | stress influence factor |
| K | Dimensionless parameter (Chapter 6) |
| k | radio between net tip resistance and OCR ratio (piezocone test) |
| m | dimensionless parameter (Chapter 6) |
| m | exponent of oedometric module equation (Chapter 7) |
| m | portion of the load supported by the granular column |
| n | drain spacing ratio not considering disturbance (Chapter 5) |
| N | gravity multiplication factor of centrifuge test |
| n | stress concentration factor (Chapter 7) |
| $n'$ | drain spacing ratio considering disturbance |
| $N_{\Delta u}$ | empirical factor of the cone in terms of pore pressure |
| $N_b$ | empirical factor of the cone in cylindrical bar test (T-bar) |
| $N_c$ | bearing capacity factor |
| $N_{kt}$ | empirical factor of the cone in terms of tip resistance |
| $N_{SPT}$ | number of blows of the SPT test |
| OCR | overconsolidation ratio |
| $q_d$ | geodrain discharge in the field – $m^3/s$ |
| $q_w$ | flow rate of the drain measured during test for a unit gradient $i = 1.0 - m^3/s$ |
| r | ratio between primary settlement ($\Delta h_a$) and total settlement ($\Delta h_a + \Delta h_{sec}$) (Chapter 4) |
| t | time – s |
| T | Time factor |
| $T^*$ | time factor (piezocone dissipation test) |
| $t_{50}$, $t_{90}$, | time $t_{100}$ required to dissipate 50%, 90% and 100% of the pore pressure, respectively – s |
| $t_{ac}$ | acceptable consolidation time according to construction schedules – s |
| $t_c$ | construction time – s |
| $t_{calc}$ | time necessary to achieve the desired consolidation – s |
| $T_h$ | time factor for horizontal drainage – s |
| $t_p$ | time corresponding to the end of primary settlement |
| $T_v$ | time factor for vertical consolidation |
| U | average degree of combined consolidation (Chapter 5) |
| U | degree of pore pressure dissipation (Chapter 2) |
| $U_h$ | average degree of horizontal consolidation (or radial) |
| $U_s$ | average degree of consolidation when surcharge is removed |
| $U_{TM}$ | average degree of consolidation according to the theory of Taylor-Merchant |
| $U_v$ | average degree of vertical consolidation |
| $W_q$ | hydraulic resistance of the PVD |
| $\alpha$ | drained strength reduction factor at soil-reinforcement interface (Chapter 6) |

$\alpha$      parameter relating OCR, undrained strength and initial effective vertical stress *in situ* (Chapter 3)

$\alpha_1$      inclination angle of the line of the graphical construction of Orleach – °

$\beta$      natural soil/treated soil settlements ratio (reduction factor of settlements)

$\beta_1$      angle of line slope of the graphical construction of Asaoka – °

$\mu$      undrained strength correction factor of vane test

$\mu_c$      parameter that combines replacement ratio and stress concentration factor in the granular column

$\mu_s$      parameter that combines replacement ratio and stress concentration factor of the soil around the granular column

$\rho$      slope of $S_u$ variation with depth – $kN/m^3$

$\Omega$      dimensionless parameter related to $\varepsilon_a$

$\alpha$      reduction factor for $S_{uo}$

## ACRONYMS

| | |
|---|---|
| C | clay |
| CAU | consolidated anisotropic undrained |
| CIU | consolidated isotropic undrained |
| $CK_oU$ | CAU tests with $K = K_o$ |
| CPT | and $CPT_u$ cone and piezocone tests, respectively |
| CRS | test carried out with constant strain rate |
| CSA | Companhia Siderúrgica do Atlântico (in Rio de Janeiro) |
| CSL | critical state line |
| CU | consolidated undrained |
| DMT | dilatometric test |
| DSS | direct simple shear test |
| E | extremely high plasticity |
| EOP | end of primary consolidation |
| FEM | finite element method |
| G.W.L. | ground water level |
| G.W.T. | ground water table |
| G.L. | ground level |
| H | high plasticity |
| M | silt |
| n.c. | normally consolidated |
| NGI | Norwegian Geotechnical Institute |
| P/60 | measurement from SPT test when with 0 blow, the sampler descends 60 cm with only the weight of the rods and sampler |
| PE | polyethylene |
| PET | polyethylene terephthalate (polyester) |
| PMT | pressiometric test |
| PP | polypropylene |
| PVA | polyvinyl acetate |
| PVC | polyvinyl chloride |
| PVD | prefabricated drains |

| | |
|---|---|
| SCPT$_u$ | piezocone seismic test |
| SDMT | seismic dilatometer test |
| SP | percussion drilling |
| T | cylindrical bar penetration test |
| UU | unconsolidated undrained |
| V | very high plasticity |
| W.T. | water table |

# Introduction

This book aims to provide the engineering student and professional with the tools necessary to understand the behavior of embankments on very soft soils. All the necessary information on how to design such earth works is provided from site investigation to monitoring and performance.

The book is based on the wide experience accumulated in the last 60 years on the design and performance of earth works on very soft to extremely soft soils in Brazil, i.e., clay soils with blows counts measured in the standard penetration test $N_{SPT}$ lower than 2. Urban settlement in Brazil took place mainly along the Brazilian coast where thick deposits of compressible soils, generally of marine and fluvial origin are found. Examples of such deposits in Southeast Brazil are the Fluminense Plains (Pacheco Silva, 1953; Almeida and Marques, 2003), Santos Plains (Massad, 2009; Pinto, 1994), the city of Recife in Northeast Brazil (Coutinho and Oliveira, 2000; Coutinho, 2007) and in South coastal areas (Dias and Moraes, 1998; Schnaid, Nacci and Militittsky, 2001; Magnani et al. 2009). Because of the extensive river system in Brazil, alluvial deposits of large thickness compressible soils also occur inland.

Chapter 1 describes the techniques used for the construction of embankments on soft soils from the most traditional to the most recent ones, comparing advantages, disadvantages and applicability of each technique.

Chapter 2 deals with *in situ* and laboratory tests used for the geotechnical site investigations necessary for the development of the geomechanical models to be used in calculations and design.

Chapter 3 presents initially, for background reference, the Cam-clay model and then describes the geotechnical properties of very soft clays illustrated by the case of a well-studied soft clay deposit in Brazil.

The theories and methods used in design are described in Chapters 4 to 7, the core of this book. Chapter 4 presents the methods used for calculating settlements and to estimate lateral displacements caused by embankment construction. The use of drains and surcharge to accelerate the settlements are described in Chapter 5. Chapter 6 discusses methods used for the stability analysis of unreinforced and reinforced embankments on soft soils. The theories and calculation methods used to design embankments on pile-like elements, i.e. piled embankments and embankments on granular piles, are described in Chapter 7.

The overall process of monitoring the performance of embankments on soft soils is described in Chapter 8 in which the more widely used instruments and interpretation methods are discussed. Monitoring is quite important as it ensures construction safety and checking of design assumptions.

Geotechnical properties of some important Brazilian soft soils are presented in the Appendix.

# Chapter 1

# Construction methods of embankments on soft soils

The most appropriate construction method to be used in a given project is associated with factors such as geotechnical characteristics of deposits, use of the area, construction deadlines and costs involved. Figure 1.1 presents some construction methods of embankments on soft soils. Some methods contemplate settlement control and others stability control, but most methods contemplate both issues. In the case of very soft soils, it is common to use geosynthetic reinforcement associated with most of the alternatives presented in Figure 1.1.

Time constraints may render inadequate techniques such as conventional embankments (Figure 1.1A,B,C,D,M) or embankments over vertical drains (Figure 1.1K,L), favoring embankments on pile-like elements (Figure 1.1F,G,H) or lightweight fills (Figure 1.1E), which, however, may have higher costs. Removal of soft soil can be used when the layer is not very thick (Figure 1.1I,J) and the transport distances are not considerable. In urban areas, it is difficult to find areas for the disposal of excavated material, considering the environmental issue associated with this disposal.

Space constraints can also prevent the use of berms (Figure 1.1B), particularly in the case of urban areas. The geometry of the embankments and the geotechnical characteristics are highly variable factors and the construction methodology must be analyzed case by case.

## 1.1 REPLACEMENT OF SOFT SOILS AND DISPLACEMENT FILLS

### 1.1.1 Replacement of soft soils

Replacement of soft soils is the partial or total removal (Figure 1.1I,J) of these soils using draglines or excavators or the direct placement of landfill to replace the soft soil. This construction method, generally used in deposits with compressible soil thicknesses of up to 4 m, has the advantage of reducing or eliminating settlements and increasing the safety factor against failure. Initially, a working platform is set up to level the terrain, just to allow the access of equipment (Figure 1.2A,B), right after the dredger starts excavating the soft soil, followed by the filling of the excavated space with fill material (Figure 1.2C,D).

Due to the very low support capacity of the top clay layers, these steps must be performed very carefully, and the equipment should be light. For very soft soils, it is noted that service roads suffer continuous settlements, as a result of the overload of

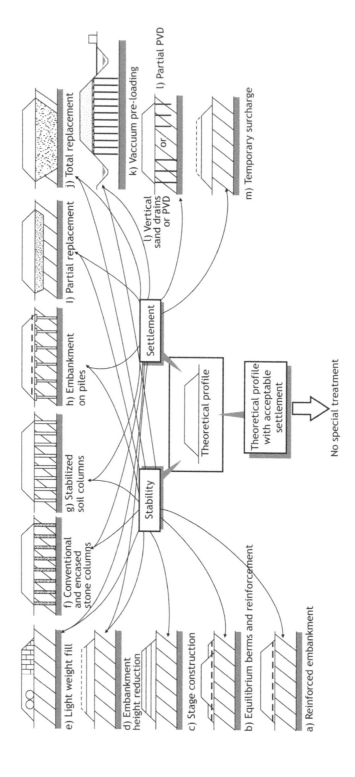

*Figure 1.1* Construction methods of embankments on soft soils (adapted from Leroueil, 1997).

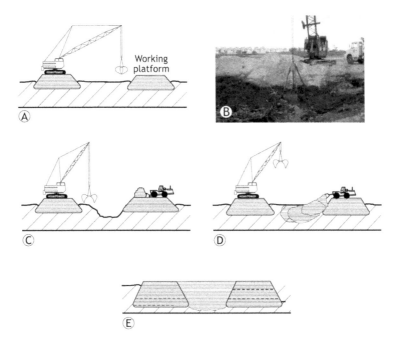

*Figure 1.2* Execution schedule for soft soil replacement: (A) and (B) excavation and removal of soft soil; (C) and (D) filling of the hole; (E) replaced soil (final condition).

equipment traffic. Shortly after, the hole is completely filled with fill material (Figure 1.2E), then, it is necessary to verify the thicknesses of the remaining soft clay through boreholes.

## 1.1.2   Displacement fills

The displacement of soft soils can be accomplished with the embankment's own weight. This technique is called displacement fill, which is the advancement of the frontal part of the embankment, which should be higher than the designed embankment. This will push and expel part of the soft soil layer, causing its rupture and leaving the embedded fill in its place (Zayen et al., 2003). The expulsion is facilitated by the lateral and frontal release of the tip fill, as shown schematically in Figure 1.3. This construction method can be used on the periphery of the area of interest by confining the internal area, allowing the embankment in this area to be constructed with a greater thicknesses.

The thickness of the remaining soft soil must be evaluated through boreholes carried out after the excavation. If there is any remaining soft soil with thickness greater than the desirable, a temporary surcharge shall be applied to eliminate post construction settlements.

One disadvantage of the replacement and displacement methods is the difficulty in quality control, because there is no guarantee that soft material will be removed evenly,

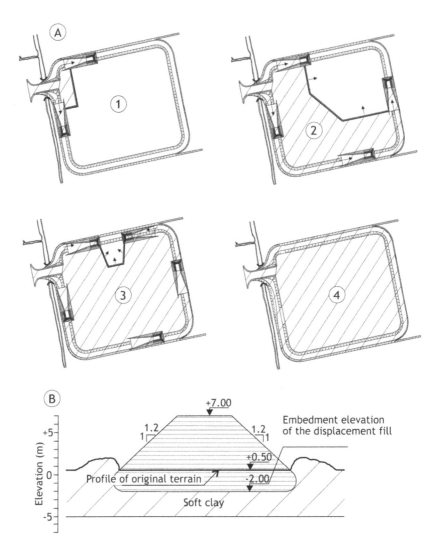

*Figure 1.3* Execution methodology of displacement fill at periphery: (A) plan; (B) cross section (Zayen et al., 2003).

which may cause differential settlements. Another disadvantage is associated with high volumes of disposed material and the difficulty of disposal, mainly in urban areas, as it is a material that cannot be reused and in certain cases may even be contaminated.

### Working platform

Working platforms, shown in Figure 1.4, are constructed to allow the access of heavy equipment in general for vertical drains installation and pile driving in case of piled embankments for instance. In some cases, the strength of the upper layer is so low that it is necessary to use constructive geotextiles reinforcement, with tensile strength

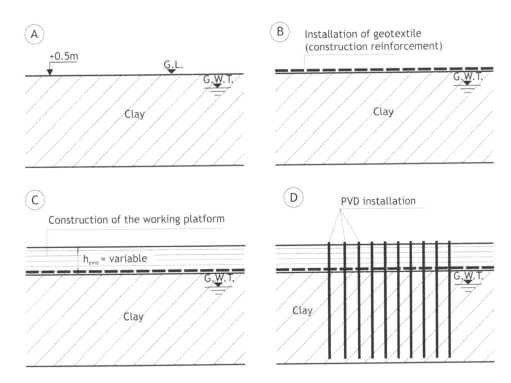

*Figure 1.4* Construction of a working platform and vertical drain installation.

between 30 kN/m and 80 kN/m to minimize the loss of fill material (Almeida et al., 2008c).

## 1.2   CONVENTIONAL EMBANKMENT WITH TEMPORARY SURCHARGE

A conventional embankment is one constructed without any specific settlement or stability control devices. The conventional embankment may be constructed with temporary surcharge (Figure 1.1M), whose function is to speed up the primary settlements and offset all or part of the secondary settlements caused by viscous phenomena not related to the dissipation of pore pressures. The temporary surcharge method is discussed in Chapter 5.

One disadvantage of this construction method is the long time necessary for settlement stabilization in low permeability very soft deposits. Therefore, one must assess the evolution of post construction settlements so that the necessary maintenance is planed.

Another disadvantage of using surcharge is the large amount of related earthworks associated. When the estimated settlements are reached, the temporary surcharge is removed and the removed material can be used as fill in another location, as described in detail in Chapter 5.

## 1.3 EMBANKMENTS BUILT IN STAGES, EMBANKMENTS WITH LATERAL BERMS AND REINFORCED EMBANKMENTS

When the undrained strength of the upper layers of soft deposit is very low, one should consider the reduction of the embankment height (Figure 1.1D). However, this reduction may not be feasible, due to requirements regarding either regional flood levels, or the geometric project of the road. In such cases, due to the low safety factor against failure, the construction of the embankment (with surcharge) may not be possible in a single stage.

The construction of the embankment in stages (Figure 1.1C), which allows the gradual gain of clay strength over time, is then a construction alternative. Stability must be verified for each stage, and for this evaluation, it is necessary to monitor the overall performance by means of geotechnical instrumentation and *in situ* tests for the necessary adjustments to the project. The increase of the clay undrained strength previously estimated in the design phase should then be verified through vane tests carried out before performing each construction stage. Construction in stages is discussed in Chapters 4 and 6.

The use of equilibrium berms (Figure 1.1B) is another solution that can be adopted to increase the safety factor ($F_s$) regarding failure. When there are restrictions as to the length of berms, or to reduce the amount of earthworks, a basal reinforcement (e.g., Magnani et al., 2009, 2010) may be installed (Figure 1.1A) with the goal of increasing the $F_s$ and better distributing stresses. These two solutions to increase the $F_s$ are addressed in the Chapter 6. The geosynthetic reinforcement must be installed after the installation of the vertical drains to avoid mechanical damage to the reinforcement.

## 1.4 EMBANKMENT ON VERTICAL DRAINS

The early vertical drains used were sand drains, which were subsequently replaced by Prefabricated vertical drains, (PVDs). The PVDs consist of a plastic core with channel-shaped grooves, encased in a low weight nonwoven geosynthetic filter, as shown in Figure 1.5A.

The drainage blanket of embankments over PVDs, is initially constructed, which also functions as a working platform (Figure 1.4), followed by the PVD installation and the construction of the embankment. In the driving process, the PVD is attached to a driving footing, which ensures that the end of the PVD is well fixed at the bottom of the layer, when the mandrel is removed (Figure 1.5B). In general, PVDs are used in association with temporary surcharge. The installation of the PVDs is carried out using driving equipment with great productivity – about 2km per day, depending on the stratigraphy – if compared to the necessary operations to install sand drains, with important financial impacts. The experience in the west part of Rio de Janeiro has an average productivity of 1km to 2km long of PVDs installed per day, for local conditions (Sandroni, 2006b).

Vacuum preloading (Figure 1.1 K) consists of the concomitant use of surcharge techniques (Figure 1.1 M) and drains (Figure 1.1L), i.e. a system of vertical (and horizontal) drains is installed and vacuum is applied, which has a preloading effect (hydrostatic). The use of PVDs and vacuum preloading are addressed in Chapter 5.

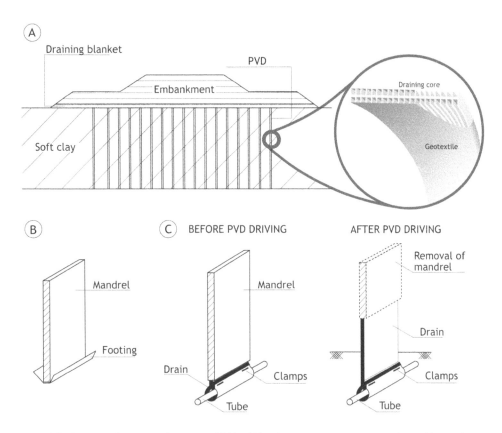

Figure 1.5 Scheme of an embankment on PVDs: (A) schematic cross section with equilibrium berms; (B) detail of the anchoring mandrel and footing of PVD; (C) detail of the driving mandrel and anchoring tube of PVDs.

## 1.5  LIGHTWEIGHT FILLS

The magnitude of primary settlements of the embankments on layers of soft soils is a function of increased vertical stress caused by the embankment built on the soft soil layer. Therefore, the use of lightweight materials in the embankment reduces the magnitude of these settlements. This technique, known as lightweight fill (Fig 1.1E), has the additional advantage of improving stability conditions of these embankments, also allowing for faster execution of the work and lessening differential settlements.

In Table 1.1 specific weights of certain materials are presented. These materials introduce voids into the embankments and are considered lightweight materials, such as, for example, expanded polystyrene (EPS), concrete pipes/galleries, etc.

Among the listed materials, EPS has been the most used (van Dorp, 1996), because when compared to other materials, it has a smaller specific weight (0.15 to 0.30 kN/m$^3$) and combines high resistance (70 to 250 kPa) with low compressibility (elastic modulus of 1 to 11 MPa). There are EPS with different weights and, strength, and when choosing an EPS, one must take into account the use of the embankment and the mobile

*Table 1.1*  Specific weights of lightweight materials for embankments.

| Material | Specific weight (kN/m³) |
|---|---|
| Expanded polystyrene – EPS (foam or similar) | 0.15 to 0.30 |
| Concrete pipes (diameter: 1 m to 1.5 m; wall thickness: 6 to 10 cm) | 2 to 4 |
| Shredded tires | 4 to 6 |
| Expanded clay | 5 to 10 |
| Sawdust | 8 to 10 |

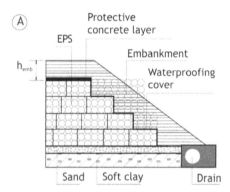

*Figure 1.6*  Use of EPS on embankments on soft soils: (A) cross section of an embankment built with EPS; (B) detail of the construction of an EPS embankment (Lima and Almeida, 2009).

loads. Figure 1.6 gives an example of lightweight fill, where the EPS core is surrounded with actual fill material with greater weight. In addition to the embankment, a protective concrete layer may be built, i.e. a slab approximately 10 cm to 15 cm thick on the lightweight fill, to redistribute stresses on the EPS, avoiding the punching of this material, caused mainly by vehicular traffic. Considering the load of the surrounding embankment and slab, preloading of the soft soil shall be done, with the use of vertical drains (usually partially penetrating) during the necessary period. The EPS may be sensitive to the action of organic solvents, thus it must be protected by a waterproofing cover insensitive to these liquids, as indicated in Figure 1.6A.

The thickness $h_{emb}$ indicated in Figure 1.6 depends on the applied loads, i.e. the use of the area. On low traffic and low load sites, this thickness will be smaller than in high traffic areas.

If the area of the embankment with EPS is subject to flooding, the EPS may float, compromising the stability and overall behavior of the embankment. In this case, the EPS base should be installed above the maximum predicted water level.

The lightweight embankment with EPS may have several formats, depending on its usage, with typical block dimensions of $4.00 \times 1.25 \times 1.00$ m, but it is possible to use blocks with different dimensions according to the demands of each project, or it is even possible to specifically cut the blocks on the worksite (Figure 1.6B). The high cost of EPS may render their implementation unsuitable in areas distant from the EPS factory, due to the cost of transporting large volumes of EPS required for the embankments.

## 1.6 EMBANKMENTS ON PILE-LIKE ELEMENTS

Embankments on pile-like elements (Figure 1.1F,G,H) are those in which all or part of the load of the embankment is transmitted to the more competent foundation soil, underlying the soft deposit and will be addressed in Chapter 7.

Embankments can be supported on piles or columns made of different materials. The stress distribution from the embankment to the piles or columns is done by means of a platform with caps, geogrids or slabs. Embankments on pile-like elements minimizes or even – depending on the adopted solution – eliminates settlements, in addition to improving the stability of embankment. One advantage of this construction method is reducing the construction schedule of the embankment, since its construction may be done in one stage, in a relatively short period.

The treatment of soft soil with granular columns (Figure 1.1F), in addition to producing less horizontal and vertical displacements when compared to conventional embankments or embankments on drains, also dissipates pore pressures through radial drainage, which speeds up the settlements and increases shear resistance of the foundation soil mass. The encasement of these columns using tubular geosynthetics with high modulus maximizes their performance.

Piled embankment (Figure 1.1H) uses the arching effect (Terzaghi, 1943), therefore allowing the stresses of the embankment to be distributed to the piles. The efficiency of the arching increases as the height of the embankment increases, consequently distributing the load to the caps and the piles (Hewlett and Randolph, 1988). Currently, geogrids are used on the caps to increase the spacing between piles.

## 1.7 CONSTRUCTION METHODOLOGIES FOR HARBOR WORKS

Soft soil deposits are common in harbor works, which are usually located in coastal areas, because of the amount of sediments that occur over thousands of years, or even recent sediment deposits, due to anthropogenic activities. In Brazil, examples of such areas are, among others, ports of Santos (Ramos and Niyama, 1994), Sepetiba (Almeida et al., 1999), Itaguaí (Marques et al., 2008), Suape (Oliveira, 2006), Itajaí-Navegantes (Marques and Lacerda, 2004), Natal (Mello, Schnaid and Gaspari, 2002), Rio Grande (Dias, 2001), and also in port areas in the Amazon region (Alencar Jr. et al., 2001; Marques, Oliveira and Souza, 2008).

Harbor works (Mason, 1982) consist essentially of an anchoring dock with a yard for holding containers in general. Figure 1.7 shows possible construction schemes for port works (Mason, 1982; Tschebotarioff, 1973). The quay is usually a structure supported by piles, which can either have an associated retaining structure or not. Examples of quays with frontal retaining structures are indicated in Figure 1.7A,B,C. The case shown in Figure 1.7A includes a relief platform. This procedure has the advantage of decreasing the active pressures on the retaining structure. In the case

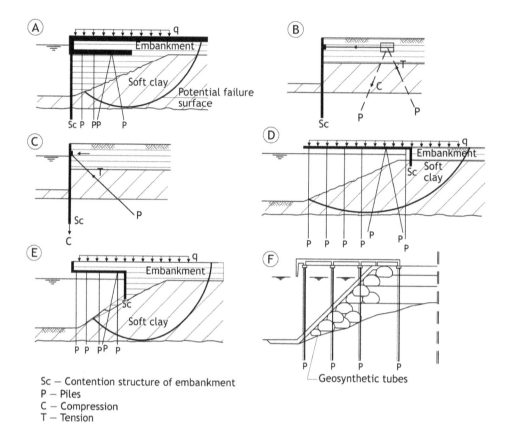

Sc – Contention structure of embankment
P – Piles
C – Compression
T – Tension

Figure 1.7 Details of geotechnical solutions in port areas.

shown in Figure 1.7 B, the retaining structure is supported by a system of two inclined piles, one being compressed and the other being tensioned. In Figure 1.7C, the retaining structure is supported by inclined piles working in tension. The compression stresses are then transmitted to the retaining structure.

In modern harbor works, which handle large vessels (the current dredging requirements reach depths of about 20 m), the retaining structures must reach great depths, so as to have the appropriate depth of embedment, in particular in the case of very thick compressible layers. Consequently, the previously described retaining structures have high costs, and alternatives have been proposed as indicated in Figure 1.7D,E. In the case shown in Figure 1.7D, the quay was expanded, and in the case of Figure 1.7E, a relief platform was used.

Figure 1.7F is a variant of Figure 1.7E and consists of an embankment, traditionally constructed with rock-fill, on the interface with the yard. An alternative to the rock-fill is the use of geotextile tubes filled with granular material or with soil cement.

Stability and settlement analysis should be carried out, regardless of the option adopted among the cases described here, and potential critical failure surfaces are shown in Figure 1.7A,D,E. In harbor works, the typical container surcharge is in the

**PHASE 1**

1 - PVD installation
2 - Area preparation: working platform geosynthetics and draining blanket
3 - Placing and installation of geosynthetic tubes
4 - Partial construction of the dikes
5 - Filling of geosynthetic tubes with the contaminated sediment to be treated

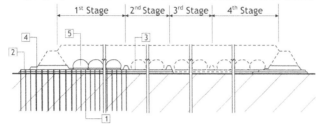

**PHASE 2**

6 - Continuing construction of dikes
7 - Construction of embankment
8 - Overload placement to accelerate settlement

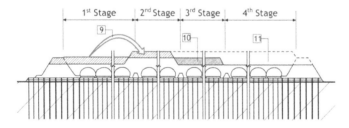

**PHASE 3**

9 - Removal of surcharge on $1^{st}$ stage and placement on $2^{nd}$ stage
10 - Displacement of surcharge from one stage to the next
11 - Repeat process until all areas have been overloaded

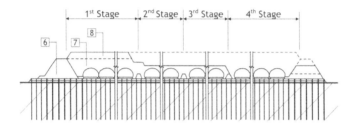

**PHASE 4**

12 - Completed and stabilized embankment

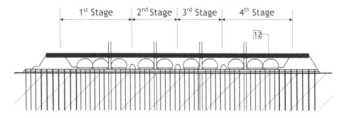

*Figure 1.8* Detail of the methodology for disposal of confined sediment.

order of 50 to 80 kPa, and the magnitude of the allowed post construction settlements will depend on technical and operational factors.

In general, harbor works require dredging thick layers of sediment. In these cases, it is common that the superficial layers present such a contamination level, that environmental agencies will not allow the disposal into water bodies. The alternative has been to dispose the sediments on land and on the harbor work site or offshore. One solution is the disposal of these dredged sediments in geotextile tubes (Leshchinsky et al., 1996; Pilarczyk, 2000), which allow the dehydration of the sediment. Also, by means of physical-chemical processes, the contaminant gets attached to the sediment and the dried fluid is then disposed of under environmentally controlled conditions.

Figure 1.8 presents a constructive scheme adopted for the disposal of such contaminated sediments in confined areas on land, which has 4 phases with 12 constructive steps explained in the figure. In some cases, the geosynthetic tubes are stacked in two or three layers. Once the landfill is completed, it can then be used as a storage area.

## 1.8   FINAL REMARKS

The planned use of the area has an important influence on the most appropriate constructive technique of embankment on soft clay. For example, on embankments of port yards, the owner may accept post construction settlements and prefer to make periodic

*Chart 1.1* Summary of construction methodologies and their characteristics.

| Construction methodologies | Characteristics |
|---|---|
| Total or partial removal of soft layer | Effective, fast, high environmental impact; boreholes are necessary for measuring the quantity of removed/remaining soil |
| Soil expulsion with controlled rupture (displacement fill) | Used for deposits of small thickness and very dependent on local experience; boreholes are necessary to gauge the thickness of the removed/remaining soil |
| Conventional embankment | Stabilization of settlements is slow |
| Construction in stages | Used, in most cases, with vertical drains; it is necessary to monitor the clay strength gain; it is not favorable for short deadlines |
| Vertical drains and surcharge | Used to accelerate settlements, large accumulated experience. Temporary surcharge may minimize or surpress secondary settlements |
| Berms and/or reinforcement | Frequently adopted; it is necessary to assess whether the tensile strength of the reinforcement is actually mobilized *in situ* |
| Use of lightweight materials | Ideal for tight deadlines; relatively high costs; its use has increased |
| Embankments on piles with geogrid platform | Ideal for tight deadlines; various layouts and materials can be used |
| Granular columns (granular piles) | Granular columns that may or may not be encased with geotextile; settlements are accelerated due to the draining nature of granular columns; geogrids are sometimes installed above the granular columns |
| Vacuum preloading | Can partially substitute the need for surcharge with fill material; horizontal strains are much smaller than those of conventional surcharge |

maintenance on the embankment, rather than investing initially on the stabilization of the settlements. However, for real estate, post construction settlements are inacceptable, since the constructor will not return to the site. On highways, settlements on bridge approaches reduce the comfort and safety of users, and on railroads, the post construction settlements should be small to minimize high maintenance costs, mainly related to the disruption of traffic. In the case of high-speed trains, for example, post construction settlements should be null.

Chart 1.1 summarizes the constructive methodologies presented in this chapter and their main features. For very soft soils it is common to use several construction techniques in parallel. For example, in the southeastern region of Brazil, particularly in the port of Santos area and in the West zone of Rio de Janeiro, in some cases, the choice has been to adopt reinforced embankments constructed in stages on vertical drains with berms and surcharge (Almeida et al., 2008c).

The decision for one executive methodology in detriment to another is a function of the geotechnical characteristics of the deposits, the use of the area (including the neighborhood), construction deadlines and the costs involved.

# Chapter 2

# Site investigation

The first stage of geotechnical work involves the planning and execution of a site investigation. Planning begins with the initial recognition of the deposit by means of geological and pedological maps, aerial photos and collection of data from previous investigations conducted in areas nearby. The next stages involve the execution of preliminary and complementary investigations. Preliminary investigations aim mainly at determining the stratigraphy of the area, and during this stage, borings are carried out. Geophysical methods are suitable tools for the evaluation of stratigraphic profiles of large areas, but they are still not used very often in soft soil investigation. At a later stage, laboratory and field investigations are carried out. The goal is to define the geotechnical parameters and the geomechanical model of the soft soil deposit, with the aim of obtaining stability and settlement calculations. The stratigraphic profile can also be obtained in this stage by means of piezocone tests.

## 2.1 PRELIMINARY INVESTIGATIONS

### 2.1.1 Borings

The preliminary investigation is the first step of the investigation itself. It consists essentially of percussion borings performed to define soil types and to estimate the thicknesses of soil layers and geological-geotechnical profiles. In very soft to soft soils, the number of blows for the last 0.30 m of the Raymond sampler is typically equal to zero ($N_{SPT} = 0$). In this case, it is possible for the sampler to penetrate 1m or more into to the soft soil if the driller does not retain the rods, or loss of the rods can also occur in the case of thick layers of soft soils. Therefore, the usual procedure is to retain the rods every 1 meter.

The main information in this phase of the investigation is the definition of the thickness of the soft clay layers, superficial fills, intermediate layers with different characteristics of the underlying soil. The borehole must be performed from a few meters into the underlying stiffer layer to characterize whether the layer is draining or not, or, alternatively, reach refusal in the case of embankments on piles. In cases of a predominat upper soft layer curves with the same thickness of soft soil layers (iso-thickness curves), as illustrated in Figure 2.1, are very useful at this stage for deciding on the construction methods that will be adopted in a given area. Geological-geotechnical profiles are also carried out, as shown in Figure 2.2.

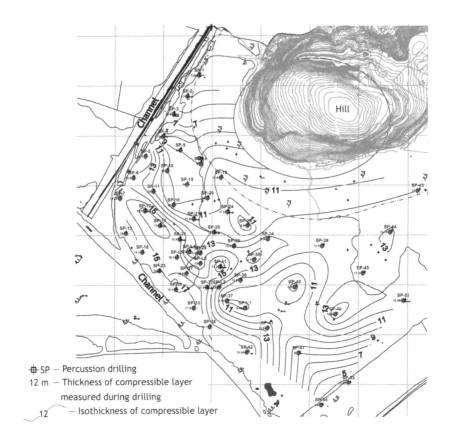

*Figure 2.1* Soft soil iso-thickness curves of a deposit in Rio de Janeiro (RJ).

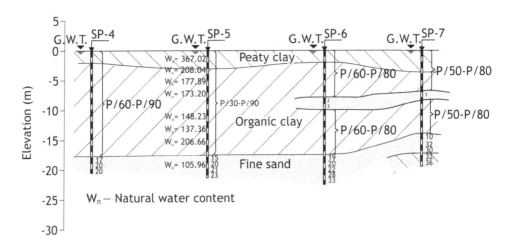

*Figure 2.2* Geological-geotechnical profile of a deposit in Rio de Janeiro (RJ).

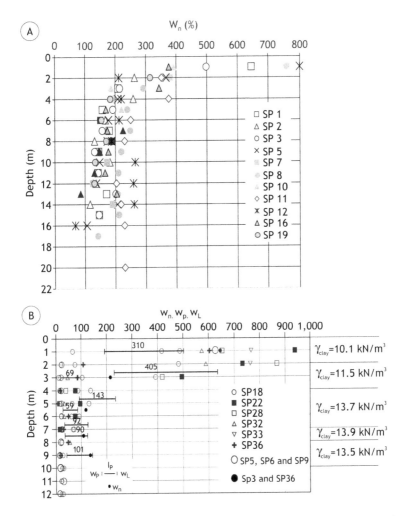

*Figure 2.3* Natural water content profiles: (A) of a deposit in Barra da Tijuca (RJ); (B) Atterberg limits and bulk unit weight of a deposit in Recreio dos Bandeirantes (Rio de Janeiro).

It is very important that the boreholes are located by coordinates and that the elevation of the borehole be recorded.

## 2.1.2 Characterization

Still in this preliminary stage, it is common to measure the natural water content ($w_n$) (Figure 2.3A) and the Atterberg limits (ASTM D4318-10) in samples taken from the SPT sampler (Figure 2.3B), since the information $N_{SPT} = 0$ in the entire layer is limited and does not distinguish between the different types of soft soils.

Measuring the water content is not costly and is needed in correlations to estimate soil parameters. To measure the water content, the sample is collected at the last

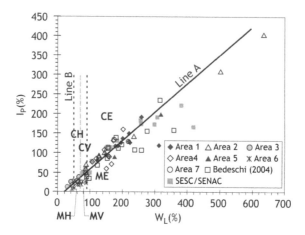

*Figure 2.4* Variation of $I_P$ with the liquid limit for clays in the west region of Rio de Janeiro (Nascimento, 2009).

15 cm of the SPT sampler such that it is not influenced by the advancement procedure often performed with percussion and water. In addition, the sample should be immediately placed in a plastic bag and stored in a Styrofoam box protected from the sunlight.

Index tests allow for qualitative evaluation of clay compressibility, when comparing $I_P$ (plasticity index) values with $w_L$ (liquid limit), as shown in Figure 2.4. In this figure, the values of $w_L$ above the B line represent high compressibility materials, named CH (high plasticity) for the range 50% < $w_L$ < 70%; CV (very high plasticity) for 70% < $w_L$ < 90% and CE (extremely high plasticity) for $w_L$ > 90% (BS 5930-BSI, 1999). According to this classification, the clays or clayey silt of the western zone in the city of Rio de Janeiro have high to extremely high plasticity. As clay soils are often organic, it is important to note that tests for determining $w_L$ and $w_P$ (plasticity limit) must be carried out without prior drying, to determine the $I_P$ later, and that the values of density of the grains $G_s$ (ABNT, 1984) of these soils are usually less than 2.6.

## 2.2 COMPLEMENTARY INVESTIGATIONS

Complementary investigations are carried out to obtain geotechnical parameters after identifying the layers. This consists of laboratory and *in situ* tests. The advantages and disadvantages of laboratory and *in situ* tests are shown in Chart 2.1. It is important to note that the strain and rupture development as well as stress paths in both *in situ* and laboratory tests differ from those *in situ* and should be considered in predictions of settlements and stability analysis.

Chart 2.2 shows the tests usually performed and the parameters estimated. As can be seen in Charts 2.1 and 2.2, laboratory and *in situ* tests are complementary. Thus,

Chart 2.1 Advantages and disadvantages of *in situ* and laboratory tests applied to soft clays (Almeida, 1996).

| Type of test | Advantages | Disadvantages |
|---|---|---|
| Laboratory | Well-defined boundary conditions | Disturbance in soils during sampling and molding |
| | Controlled drainage conditions | Low representation of tested soil |
| | Known stress paths during test | Under similar conditions it is generally more expensive than field tests |
| | Identifiable nature of the soil | |
| *In situ* | Soil tested in its natural environment | Poorly defined boundary conditions, except for self-boring pressuremeter |
| | Continuous measurement with depth (CPT, piezocone) | Unknown drainage conditions |
| | Greater volume of tested soil | Unknown degree of disturbance |
| | Usually faster than laboratory test | Non-identified nature of soil (except percussion boreholes) |

it is common to conduct a cluster of boreholes in contiguous verticals (about 2 m apart), including laboratory and *in situ* tests, as presented at the end of the chapter.

### 2.2.1 *In situ* tests

During this phase, the most commonly used *in situ* tests are the vane and the piezocone tests (Schnaid, 2009). Other *in situ* tests (Danziger and Schnaid, 2000; Coutinho, 2008) conducted on soft deposits are the dilatometer tests (e.g., Soares, Almeida and Danziger, 1987) and the Tbar test (Stewart and Randolph, 1991; Almeida, Danziger and Macedo, 2006). The latter, although mostly used in offshore research, has the potential to be used onshore, due to its simplicity, since no pore pressure measurement is necessary.

Seismic piezocone tests (SCPTu) or seismic dilatometer tests (SDMT) allow determination of the small-strain shear modulus the $G_o$ (or $G_{max}$), which can be correlated with undrained Young's modulus ($E_u$). This "elastic" parameter is less relevant in the case of embankments on soft soils, typically with relatively low safety factors ($Fs \approx 1.5$).

### 2.2.2 Laboratory tests

The laboratory tests usually performed for the design of embankments on soft soil are the ones for the determination of the index properties (grain size analysis; liquid and plasticity limits; specific gravity of the grains), as well as oedometer and triaxial tests. In some cases, the amount of organic matter is determined by weight. One can use the measure of the weight loss in the oven with temperature above 440°C (NBR 13600 – ABNT, 1996), a procedure which is quicker and less expensive or, preferably, a method (e.g., Embrapa, 1997), by determining the organic carbon content.

*Chart 2.2* General characteristics of *in situ* and laboratory tests, estimated geotechnical parameters and recommendations.

| Test | Type | Aim of the test | Main parameters estimated | Other parameters | Notes and recommendations |
|---|---|---|---|---|---|
| Laboratory | Index tests | General characterization of the soil; interpretation of other tests | $w_n$, $w_L$, $w_p$, $G_s$, grain size distribution curves | Compressibility estimates | Recommended to assess the organic matter in very organic soils and peat |
| | Oedometer consolidation test | Calculation of settlements and settlements vs. time | $C_c$, $C_s$, $\sigma'_{vm}$, $c_v$, $e_o$ | $E_{oed}$, $c_\alpha$ | Essential for calculating the magnitude and rate of settlements; can be replaced by CRS test |
| | Triaxial UU | Stability calculations ($S_u$ is affected by disturbance) | $S_u$ | | More affected by disturbance than CU test |
| | Triaxial CU | Stability calculations; parameters for deformability calculations 2D (FEM) | $S_u$, $c'$, $\phi'$ | $E_u$ | CAU test (anisotropic consolidation) is more indicated |
| *In situ* | Vane | Stability calculations | $S_u$, $S_t$ | OCR | Essential for determining the clay undrained strength |
| | Piezocone (CPTu) | Stratigraphy; settlements vs. time (dissipation test) | Estimation of $S_u$, $c_h$ ($c_v$) profile | OCR profile, $K_0$, $E_{oed}$ | Highly recommended; low cost/benefit ratio |
| | Tbar | Undrained strength | Estimation of $S_u$, profile | | Does not require pore pressure correction; most commonly used offshore |
| | Dilatometer (DMT) | complementary test, general | $S_u$, OCR, $K_0$ | $c_h$, $E_{oed}$ | – |
| | Pressuremeter (PMT) | complementary test, general | $S_u$, $G_o$ | $c_h$ | – |

## 2.3  VANE TESTS

### 2.3.1  Equipment and procedures

The vane test is the most used tool to estimate undrained strength ($S_u$) of soft soils. The test consists of rotating a vane blade at a constant rotation of 6° per minute at predefined depths. The maximum value of the torque measured is then used to obtain the clay undrained shear strength ($S_u$). The $S_u$ value is affected by the following factors: mechanical friction, vane blade characteristics, vane rotation rate, clay plasticity, disturbance, heterogeneity and clay anisotropy. The calculated value of $S_u$ is influenced by the rupture hypothesis adopted (Chandler, 1988). Accordingly, several precautions are advisable when performing this test, which are outlined in NBR 10905 (ABNT, 1989). For example, the necessary standardization of waiting time between the installation and rotation of the vane is fixed as five minutes, so that the value of $S_u$ is not over-estimated due to the drainage that can occur due to longer wait periods. The rate of rotation, the dimensions of the vane and the test time should also be specified in the standard adopted.

This test should be ideally carried out by measuring the torque near the vane, otherwise the friction in the rods is included in the measurement and should be corrected. Moreover, the rotation angle – generally measured on the surface of the terrain – incorporates the elastic rotation of the vane rod, which is high, in the case of larger depths.

An equipment with torque measurement near the vane blade was developed through a partnership between COPPE/UFRJ and UFPE (Almeida, 1996; Nascimento, 1998; Oliveira, 2000), that is equipped with protection shoe where the vane blades are lodged (Figure 2.5). These types of equipment have since then been used with excellent results (e.g., Crespo Neto, 2004; Jannuzzi, 2009; Baroni, 2010).

### 2.3.2  Undrained strength

The measurement of torque (T) versus rotation in the vane test allows the determination of the values of the undrained strength ($S_u$) of the natural and disturbed soil. The usual hypotheses adopted for the calculation of $S_u$ are: undrained conditions, isotropic soil and constant strength around the vane. With these assumptions, and for a height to diameter ratio (H/D) equal to 2, the equation used for calculating $S_u$ based on the maximum measured torque value, outlined by NBR 10905 (ABNT, 1989), is:

$$S_u = \frac{0.86T}{\pi D^3} \tag{2.1}$$

Wroth (1984) showed experimental results which demonstrated that the hypothesis of constant $S_u$ at the top and at the base of the vane is not valid. Consequently, based on studies carried out with the London clay, Eq. (2.1) may provide, in theory, conservative results of the order of 9%. Schnaid (2009) shows theoretical equation proposed in the literature considering different rupture modes.

Eq. (2.1) is also used to calculate the disturbed strength of clay ($S_{ur}$), a measure that consists of rotating the vane 10 complete turns after reaching the maximum torque,

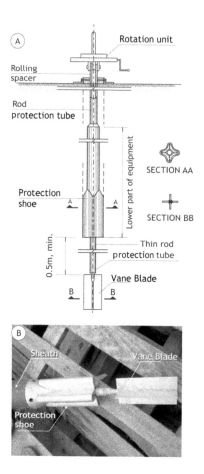

Figure 2.5 Vane Equipment: (A) equipment components; (B) detail of protection shoe.

so as to remold the soil and then measure the disturbed strength. It is recommended that the interval between the two phases of the test be less than 5 min.

Tests on undisturbed natural soil must result in moderate rotation angles for peak values (Figure 2.6). Baroni (2010) noticed a variation of 5° to 25°, with some isolated points (peat, shell lenses) where $\theta_{max}$ reached 56°. The average rotation angle corresponding to the maximum torque applied on three deposits of Barra da Tijuca (Baroni, 2010) was 16°. The quality of the vane test can be assessed by the shape of the torque curve versus rotation angle . In general, peak rotation angles greater than 30° indicate some clay disturbance.

### 2.3.3   Clay sensitivity

Figure 2.7 shows examples of undrained strength measured in vane tests, with test results for natural ($S_u$) and remolded ($S_{ur}$) conditions. The sensitivity $S_t$ of a clay is

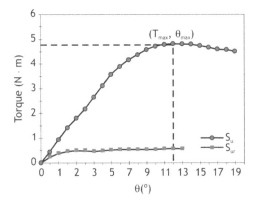

Figure 2.6 Torque versus rotation angle for tests on natural and disturbed clay (Crespo Neto, 2004).

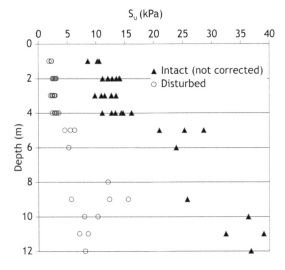

Figure 2.7 $S_u$ natural and disturbed profile vs. depth – Sarapuí II Clay (Jannuzzi, 2009).

defined as the ratio of the peak strength ($S_u$) to the remolded strength ($S_{ur}$), according to the equation:

$$S_t = \frac{S_u}{S_{ur}} \qquad (2.2)$$

The classification of clays according to sensitivity appears in Table 2.1 (Mitchell, 1993).

Most Brazilian clays have sensitivity in the 1 to 8 range, with average values between 3 and 5 (Schnaid, 2009). However, sensitivity values of up to 10 have been observed in clays of Rio de Janeiro, such as the clays of Juturnaíba (Coutinho, 1986) and of Barra da Tijuca (Macedo, 2004; Baroni, 2010).

*Table 2.1*   Classification of clays according to sensitivity (Mitchell, 1993).

| Soil type | $S_t$ (sensitivity) |
| --- | --- |
| Insensitive clays | 1 |
| Slightly sensitive clays | 1–2 |
| Medium sensitive clays | 2–4 |
| Sensitive clays | 4–8 |
| Slightly quick clays | 8–16 |
| Medium quick clays | 16–32 |
| Very quick clays | 32–64 |
| Extra quick clays | >64 |

### 2.3.4   Stress history

Stress history is commonly expressed by the overconsolidation ratio OCR $= \sigma'_{vm}/\sigma'_{vo}$, where the overconsolidation stress $\sigma'_{vm}$ is determined through the oedometer test and the effective vertical stress $\sigma'_{vo}$ *in situ* is determined through geotechnical profiles. Since good quality samples are hard to obtain in very soft clays, overconsolidation stress values $\sigma'_{vm}$ are not always reliable. $\sigma'_{vo}$ values may also be susceptible to errors, particularly in the upper layers, due to low $\sigma'_{vo}$ values resulting from difficulties in estimating the water level and the exact position of the sample in depth within the sampler. In addition, total unit weight values lower than $12\,kN/m^3$ are not uncommon in extremely soft organic clays, as shown in Figure 2.3B.

As a result of the issues discussed here, it is common to use *in situ* tests to estimate the OCR values of the clay. Among the *in situ* tests, the vane test can be used for this purpose. In this case, it is possible to use the equation proposed by Mayne and Mitchell (1988):

$$OCR = \alpha \frac{S_u}{\sigma'_{vo}} \tag{2.3}$$

where the value of $\alpha$ can be obtained from the correlation with the plasticity index, given by:

$$\alpha = 22 \cdot (I_P)^{-0.48} \tag{2.4}$$

Another way to estimate the OCR value is from the $S_u/\sigma'_{vm}$ versus $I_p$ relationship, as shown in Figure 2.8, for deposits of different origins. In fact, Brazilian clays have very high plasticity, unlike clays of Eastern Canada (Leroueil, Tavenas and Le Bihan, 1983; Marques, 2001). The piezocone test, described in section 2.4, has been more frequently used for estimating the overconsolidation stress than the vane test.

### 2.3.5   Clay anisotropy

The anisotropy originates from the mode of deposition of the clay (inherent anisotropy) and from induced strains after deposition (induced anisotropy). Strength anisotropy studies with vane tests have been carried out (Aas, 1965; Collet, 1978) with different H/D vane blades ratio, aiming to measure clay strength in the horizontal $S_{uh}$ and

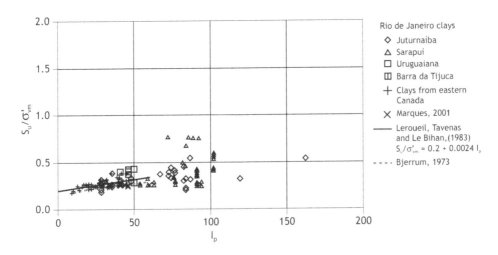

*Figure 2.8* Variation of the normalized undrained strength ratio $S_u/\sigma'_{vm}$ with the plasticity index $I_p$.

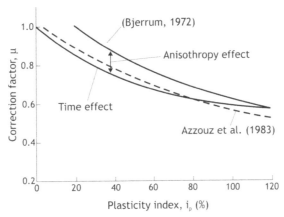

*Figure 2.9* Correction factor of ($S_u$) values of the vane test related to plasticity index (Bjerrum, 1972).

vertical $S_{uv}$ directions to obtain the strength anisotropy ratio $S_{uh}/S_{uv}$. These studies indicate (Bjerrum, 1973) that the $S_{uh}/S_{uv}$ ratio is close to unity for slightly very soft overconsolidated clays, to soft clays with plasticity indexes over 40%.

## 2.3.6  Test correction

The undrained stress ($S_u$) measured in the vane test should be multiplied by a correction factor (Bjerrum, 1972) to obtain the design strength:

$$S_u \text{ (design)} = \mu S_u \text{ (vane)} \qquad (2.5)$$

The correction factor $\mu$, shown in Figure 2.9, is a function of the clay's plasticity index and incorporates two effects: clay anisotropy and the difference between the loading rate in the field and the vane test rate (time effect).

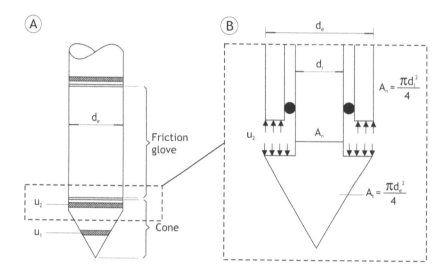

*Figure 2.10* Details of the piezocone probe: (A) measure of pore pressure at two points; (B) details of the pore pressure acting on the tip.

## 2.4 PIEZOCONE TEST

### 2.4.1 Equipment and procedures

The piezocone test consists of continuous penetration – at constant rate of penetration of the order of 2 cm/s – of a cylindrical element with a cone-shaped tip with continuous measurement of tip strength ($q_c$), lateral friction ($f_s$) and pore pressure (u), as shown in Figure 2.10A (Lunne, Robertson and Powell, 1997; Schnaid, 2008). The standardization of the penetration rate is important, since the strength value varies by about 10% per logarithmic cycle of the penetration rate (Leroueil and Marques, 1996; Crespo Neto, 2004).

It is ideal to measure pore pressure at two points: one on the face ($u_1$) of the cone and another at the base of the cone ($u_2$). However, the majority of the equipment can only measure $u_2$, which is required for the correction of the measured cone resistance.

The CPTu probe used in soft soils, generally has an area of 10 cm², but probes with smaller areas are also used with the aim of speeding up the dissipation tests (Baroni, 2010). The CPTu equipment in very soft soils should be lightweight, so as to facilitate its accessibility, especially in areas of low bearing top layers.

### 2.4.2 Correction of cone resistance

The piezocone test has been used for preliminary estimation of typical soil behavior, definition of soft soil deposit stratigraphy, definition of the continuous profile of undrained strength and compressibility parameters of the soil, in addition to other parameters, as described in chart 2.2.

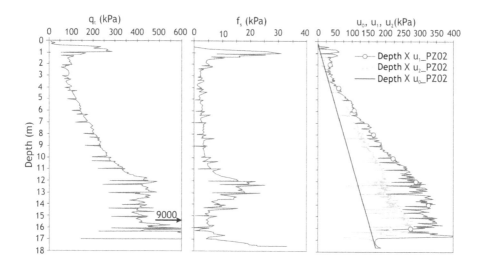

*Figure 2.11* Typical results of a vertical piezocone test conducted in Barra da Tijuca (RJ). (A) $q_t$ profile; (B) lateral friction strength profile, $f_s$; (C) pore pressure profile (Baroni, 2010).

Because the pore pressure operates unequally on the geometry of the cone tip (Figure 2.10B), the cone resistance measured at the tip of the cone ($q_c$) must be corrected according to the equation:

$$q_t = q_c + (1 - a)u_2 \tag{2.6}$$

where $q_t$ is the corrected cone resistance, $q_c$ is the cone resistance measured in the cone; $u_2$ is the pore pressure measured at the base of the cone; and a is the ratio of the areas $A_n/A_t$ (Figure 2.10B).

The geotechnical engineer should request the cone tip characteristics (radius, $A_n$, $A_t$) as well as the raw data of the test, in order to be able to interpret the results. The value of the parameter a could also be obtained by means of calibration. Figure 2.11 shows typical results of a piezocone test ($q_t$, $f_s$ and u profiles) conducted in a deposit at Barra da Tijuca (RJ).

### 2.4.3   Preliminary soil classification

Several approaches for preliminary soil classification based on piezocone test results are available in the literature. The chart proposed by Robertson (1990) (Figure 2.12) is one of the most used. With the parameters used in the chart, and defined in the figure, it is possible to obtain the estimate of soil type for each depth, in general at every 2 cm.

### 2.4.4   Undrained strength ($S_u$)

The undrained strength $S_u$ may be obtained from the piezocone test using a number of equations (Lunne, Robertson and Powell, 1997; Schnaid, 2008). The most used

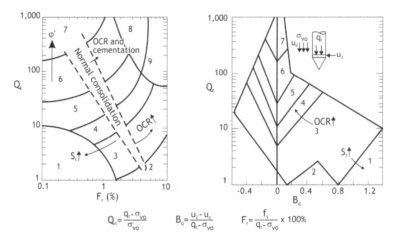

$$Q_t = \frac{q_t - \sigma_{vo}}{\sigma_{vo}} \qquad B_q = \frac{u_2 - u_o}{q_t - \sigma_{vo}} \qquad F_r = \frac{f_s}{q_t - \sigma_{vo}} \times 100\%$$

$u_o$ – Hydrostatic pore pressure at depth of dissipation test

| Zone | Soil behaviour type | Zone | Soil behaviour type | Zone | Soil behaviour type |
|------|---------------------|------|---------------------|------|---------------------|
| 1. | Sensitive, fine grained; | 4. | Silt mixtures clayey silt to silty clay | 7. | Gravelly sand to sand; |
| 2. | Organic soils-peats; | 5. | Sand mixtures; silty sand to sand silty | 8. | Very stiff sand to clayey sand |
| 3. | Clays-clay to silty clay; | 6. | Sands; clean sands to silty sands | 9. | Very stiff fine grained |

*Figure 2.12* Soil behavior type classification chart based on normalized CPTu data (Robertson, 1990).

equations relate the corrected strength ($q_t$) of the cone with the cone factor ($N_{kt}$), and the pore pressure with the cone factor ($N_{\Delta u}$), as shown below:

$$S_u = \frac{q_t - \sigma_{vo}}{N_{kt}} \qquad\qquad (2.7)$$

$$S_u = \frac{u_2 - u_0}{N_{\Delta u}} = \frac{\Delta u}{N_{\Delta u}} \qquad\qquad (2.8)$$

In geotechnical practice, Eq. (2.8) is less used than Eq. (2.7). The $N_{kt}$ value to be used in Eq. (2.7) should be obtained from the correlation of the piezocone tests and the undrained strength from vane tests which are most commonly used for this purpose. The experiences accumulated for nearly 30 years conducting piezocone tests indicate that the $N_{kt}$ value should be obtained for layers with different characteristics in the same deposit. $N_{kt}$ values are obtained for each depth and an average value for the deposit is to be used in Eq. (2.7) to obtain the estimated profile of $S_u$. Due to the heterogeneity of the deposit, the average value can vary considerably, as shown in Figure 2.13. In this case, it is possible to use a single $N_{kt}$ value for each profile or vary the value along the depth. $N_{kt}$ values vary typically between 10 and 20, and studies indicate (e.g., Ladd and De Groot, 2003) that this correlation also depends on the CPTu equipment used. Table 2.2 shows typical $N_{kt}$ values, as well as some

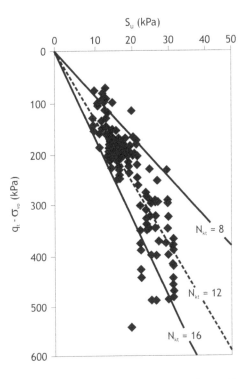

*Figure 2.13* Cone factor values, $N_{kt}$, obtained from tests in Porto Alegre (Schnaid, 2000).

parameters for Brazilian soils. The average of the $N_{kt}$ factor is about 12 for Brazilian clays (Almeida, Marques and Baroni, 2010).

## 2.4.5 Stress history

Several equations have been suggested in the literature to obtain OCR vs. depth profiles by means of piezocone tests. The one that is mostly used is:

$$OCR = k \cdot Q_t \qquad (2.9)$$

where:

$$Q_t = \frac{q_t - \sigma_{vo}}{\sigma'_{vo}} \qquad (2.10)$$

k values in the 0.15–0.50 range have been obtained in various clayey deposits (Schnaid, 2009), with the recommended value of the order of 0.30. The lowest range has been registered for Brazilian very soft clays (Jannuzzi, 2009; Baroni et al., 2012).

Table 2.2 Typical $N_{kt}$ values for Brazilian soils (Danziger and Schnaid, 2000).

| Location | Characteristics of clays | | | Average $N_{kt}$ (variable) | Notes | References | |
|---|---|---|---|---|---|---|---|
| | $I_p$ (%) | OCR | $S_u$ (kPa) | | | Characteristics of clays | Piezocone tests |
| Sarapuí, Rio de Janeiro | 100–250 | 1.3–2.5 (below the 3 m crust) | 8–18, vane | 9 (8 to 10) – 3 m to 6.5 m 10.5 (10 to 11) – 6 m to 10 m | Other piezocone tests carried out on site (below embankment): Alencar Jr. (1984); Rocha Filho and Alencar (1985) | Ortigão (1980); Ortigão, Werneck and Lacerda (1983); Ortigão and Collet (1986) | Soares et al. (1986), Soares, Almeida and Danziger (1987); Sills, Almeida and Danziger (1988); Ortigão (1990); Danziger, Almeida and Sills (1997) Bezerra (1996) |
| Senac, Rio de Janeiro | 100–50 | 1.5 (below the 3 m crust) | 8–30, vane | 14 (11 to 16) – 9 (5 to 11) – | | Almeida (1998) | |
| Clube Internacional, Recife | 25–90 | 1–2 | 35–55, UU | 12.5 (7 m to 16 m) – 13–16 m to 26 m total variation: 10 to 15.5 (11 to 17) – 7 m to 16 m (12 to 16) – 16 to 26 m | | Coutinho, Oliveira and Danziger (1993) | Oliveira (1991); Coutinho, Oliveira and Danziger (1993) Bezerra (1996) |
| Ibura, Recife | 45–115 | aprox. 1 (below the crust) | 9–27, UU | 14–4 at 11.1 m 13.5–11.1 m to 21 m | | Coutinho, Oliveira and Oliveira (1998) | |
| Port of Aracaju | 25–45 | 1–2 | 10–30, vane | 15.5 (14.5 to 16.5) | | Brugger et al. (1994); Sandroni et al. (1997) | Danziger (1990); Brugger et al. (1994); Sandroni et al. (1997) |
| Port of Santos | 40–80 | 1.3–2 | 5–50, vane | 18 (15 to 21) | | Samara et al. (1982) | Bezerra (1993); Almeida (1996) |
| Enseada do Cabrito, Salvador | 50 | 1.5–3 | 9–17, vane | 15 (12 to 18) | | Baptista and Sayão (1998) | |
| Ceasa, Porto Alegre | 20–70 | 1–1.5 | 10–20, vane | 12 (8 to 16) | | Soares, Schnaid and Bica (1994, 1997) | |
| Salgado Filho Airport, Porto Alegre | 20–70 | 1–5 | 10–15, UU-CIU | 12 (10 to 16) | | Schnaid et al. (1997) | |

*Table 2.3* Time factor T* according to degree of consolidation (U) (Houlsby and Teh, 1988).

| U (%) | Time factor T* according to the position of pore pressure transducer | |
| | Face of the cone ($u_1$) | Base of the cone ($u_2$) |
| --- | --- | --- |
| 20 | 0.014 | 0.038 |
| 30 | 0.032 | 0.078 |
| 40 | 0.063 | 0.142 |
| 50 | 0.118 | 0.245 |
| 60 | 0.226 | 0.439 |
| 70 | 0.463 | 0.804 |
| 80 | 1.040 | 1.600 |

## 2.4.6   Coefficient of consolidation

Tests with dissipation of the excess pore pressures generated during the penetration of the piezocone in the ground can be used to estimate the coefficient of horizontal consolidation ($c_h$) and, with this, estimate the coefficient of vertical consolidation ($c_v$). The test consists of interrupting the penetration of the piezocone at predefined depths, until reaching at least 50% of the dissipation of the generated excess pore pressure.

The most used $c_h$ estimation method is the one by Houlsby and Teh (1988), which considers the rigidity index of the soil ($I_R$), with the time factor defined as follows:

$$T^* = \frac{c_h \cdot t}{R^2 \sqrt{I_R}} \qquad (2.11)$$

where:
R – piezocone radius
t – dissipation time/period
$I_R$ – rigidity index ($G/S_u$)
G – shear stress modulus of the soil. Generally, $G = E_u/3$ is used, where $E_u$ is Young's undrained modulus, usually obtained from CU tests at 50% of the maximum deviatoric stress.

In Table 2.3, values for the time factor T* are listed according to the pore pressure degree of consolidation (U), observing that the solution is a function of the porous element's position in the cone.

The measurement of $u_2$ at the base of the cone is standard and is mostly used to interpret piezocone dissipation results. Any procedure to estimate $c_h$ (e.g. Robertson et al., 1992; Danziger, Almeida and Sills, 1997) requires the accurate measurement of the pore pressure of the pore pressure value at the start of dissipation $u_i$, and the hydrostatic pore pressure value $u_0$. The most common procedure (Robertson et al., 1992) is the determination of the pore pressure value $u_{50} = (u_i - u_0)/2$, corresponding to 50% dissipation. It is then possible to obtain the time $t_{50}$, as shown in Figure 2.14. However, the most accurate procedure is to obtain T* and, then $c_h$ by overlapping the experimental and theoretical curves, as proposed by Danziger, Almeida and Sills (1997).

Robertson et al. (1992) proposed a direct way to estimate of $c_h$ from the $t_{50}$ value using the chart in Figure 2.15, which was developed using Eq. (2.11) and data from

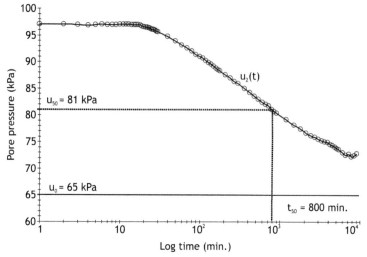

$u_i = 97$ kPa – Porepressure at the begining of dissipation test
$u_0 = 65$ kPa – Hydrostatic pore pressure at the depth of dissipation test
$\Delta u = 32$ kPa
$\Delta u_{50} = 16$ kPa
$u_{50} = (97 - 16)$ kPa $= 81$ kPa

*Figure 2.14* Example of $c_h$ calculation – dissipation test in Barra da Tijuca (RJ).

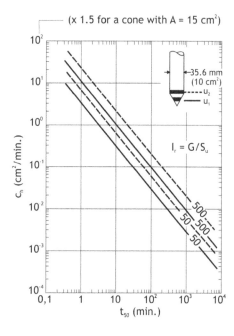

*Figure 2.15* Chart for obtaining $c_h$ from $t_{50}$ (Robertson et al., 1992).

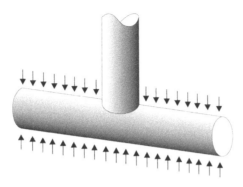

*Figure 2.16* Detail of the T-bar penetrometer.

Table 2.3. This chart is valid for $I_R$ values varying between 50 and 500 and for cone areas of $10\,cm^2$ and $15\,cm^2$.

The $c_h$ value should be corrected to account for anisotropy and stress range when calculating consolidation rate and comparing with measured $c_v$ values from oedometer tests in the normally consolidated range $c_{v(na)}$. Equations used for the conversion of $c_h$ to $c_{v(na)}$ are available in Lunne, Robertson and Powell (1997) and in Schnaid (2009).

## 2.5  T-BAR TEST

The T-bar test (Stewart and Randolph, 1991; Randolph, 2004) has been mostly used recently to obtain the undrained strength of clayey soils. This test consists of the penetration of a cylindrical bar in the ground (Figure 2.16) and has the advantage of not requiring pore pressure corrections, since there is equilibrium of the stresses acting below and above the bar. In the T-bar test, the undrained strength is given by:

$$S_u = \frac{q_b}{N_b} \qquad (2.12)$$

where $N_b$ is an empirical factor of the cone, whose theoretical value is 10.5, and $q_b$ is the strength measured during the test. Studies (Almeida, Danziger and Macedo, 2006; Long and Phoon, 2004) indicate that these values are consistent with $S_u$ values from vane tests.

## 2.6  SOIL SAMPLING FOR LABORATORY TESTS

An essential condition for good laboratory test results is the suitability of undisturbed samples. Sampling involves several operations, with stress state changes and induced disturbance of the soil, as shown in Figure 2.17. However, even perfect hypothetical sampling results in an inevitable relief in the stress state of the soil (Ladd and Lambe, 1963; Hight, 2001).

The soil sample retrieval using a Shelby sampler with a stationary piston (NBR9820 – ABNT, 1997; ASTM D1587-08) requires special precautions such as the use of bentonite slurry in the borehole. After installation of the Shelby tube into to

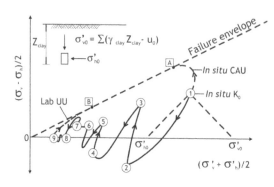

| Portion/event | Event |
|---|---|
| 1-2 | Perforation |
| 2-3-4-5 | Sampler tube installation |
| 5-6 | Removal of sampler tube |
| 6-7 | Transport and storage |
| 7-8 | Extrusion of sample from tube |
| 8-9 | Preparation of test specimen |

Figure 2.17 Variation of the stress states of a sample during sampling.

Figure 2.18 Procedure for extrusion and preparation of soft soil specimens in the laboratory: (A) cutting the sampler, (B) needle and steel wire used to separate the sample from the sampler (Baroni, 2010).

the soil, it is sometimes necessary to wait a few hours to collect the sample from the ground to minimize disturbance.

In the laboratory, the procedure proposed by Ladd and De Groot (2003) should be used for the extrusion of the samples from the Shelby sampler. This consists of cutting the sample tube to the required length for the specimen to be tested (Figure 2.18A), driving a needle of suitable length between the specimen and the sampler wall, then passing a metal wire around this interface in order to release the sample from the sampler (Figure 2.18B).

## 2.7 OEDOMETER CONSOLIDATION TESTS

The oedometer consolidation test is essential for calculating the magnitude of settlement and their evolution over time. In the conventional oedometer test with incremental load, each load increment is applied for 24 hours.

The maximum vertical stress to be applied should be chosen depending on the stress history of the deposit and the embankment height. For very soft clays, one must start with low vertical stresses of 1.5 kPa or 3 kPa, and then the load is doubled. To better determine the overconsolidation stress, sometimes intermediate loading stages are also carried out. Loading stages will be carried out until the required vertical stress, which should be in the order of 400 kPa, at least, even for embankments that are low in height. This stress level allows better definition of the virgin compression curve and allows the evaluation of the suitability of the sample, since good quality soft clay samples have a clear curve in the log scale for virgin range of stress.

Typical tests are two weeks long, particularly if an unloading cycle is carried out. Direct permeability measurement through the variable load test (Head, 1982) in some cases is also carried out for some load stages, which should lead to a longer test period since this is carried out during the 24 hours following the end of the consolidation phase of a load stage, i.e., the duration of each stage subjected to this additional test is 48 hours.

Figure 2.19A, B shows the correlation of the soil compression index with the natural moisture content ($w_n$) in some Rio de Janeiro clays and clays located in Rio de Janeiro's western region, respectively. In the preliminary design stage, this correlation allows to estimate the magnitude of settlements that will occur with the construction of the embankments.

## 2.7.1  Other consolidation tests

The constant rate of strain consolidation test (CRS) generates the compression parameters within a much shorter period than the incremental test (Wissa et al., 1971, Head, 1982). This type of test has been conducted on Rio de Janeiro clays (Almeida et al., 1995) but is less used in the current Brazilian practice.

The automated oedometer test equipment that is currently available comercially allows the load to be applied automatically in sequence, without the need to wait 24 hours for each load stage. This test, also known as the "accelerated incremental consolidation test", usually lasts about two days, the same average duration of a typical CRS test. This should be set to a given criterion for the application of each loading sequence such as, for example, the end of the primary ($t_{100}$), based on the Taylor $t_{90}$ method. The compression curve obtained with this test differs from the usual test with the 24-hour criterion, since the rates at the end of the primary strain/deformation are greater than the 24-hour rates. The overconsolidation stresses of the accelerated incremental test are greater than those from the conventional tests, and should be appropriately considered when using the results of each type of test since the correlations in the literature use OCR values obtained from conventional 24-hour oedometer tests.

## 2.7.2  Sample quality

Consolidation test results are very dependent on the sample quality. Lunne, Berre and Strandvik (1997) proposed a relatively more restrictive criterion for assessing the quality of samples than the recommendations of Coutinho (2007) and Sandroni (2006) for Brazilian clays, as indicated in Table 2.4. These recommendations are based on the $\Delta e/e_{vo}$ relationship, where $\Delta e$ is the variation of the voids from the beginning of the

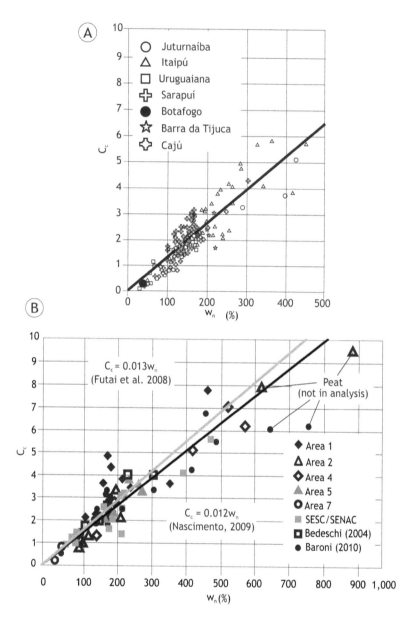

*Figure 2.19* Compression index (C$_c$) × natural moisture content (w$_n$): (A) Rio de Janeiro clays (Futai, 1999; Almeida et al., 2008), (B) clays of Barra da Tijuca and Recreio dos Bandeirantes (Almeida et al., 2008c).

test to the effective vertical stress *in situ* $\sigma'_{vo}$, and e$_{vo}$ is the void ratio corresponding to $\sigma'_{vo}$. Baroni et al. (2012) used the criterion proposed by Coutinho (2007) for Rio de Janeiro clays and noted that, despite all precautions taken during sampling, in the case of these clays, 83% of the samples were good or fair.

*Table 2.4*  Criteria for classification of sample quality.

| OCR | $\Delta e/e_{vo}$ | | | |
| | Very good to excellent | Good to fair | Bad | Very bad |
| --- | --- | --- | --- | --- |
| Lunne, Berre and Strandvik (1997) Criterion | | | | |
| 1–2 | <0.04 | 0.04–0.07 | 0.07–0.14 | >0.14 |
| 2–4 | <0.03 | 0.03–0.05 | 0.05–0.10 | >0.10 |
| Sandroni (2006b) Criterion | | | | |
| <2 | <0.03 | 0.03–0.05 | 0.05–0.10 | >0.10 |
| Coutinho (2007) Criterion | | | | |
| 1–2.5 | <0.05 | 0.05–0.08 | 0.08–0.14 | >0.14 |

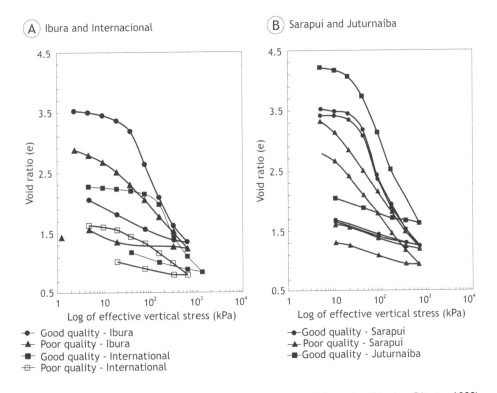

*Figure 2.20*  $e \times \log \sigma'_v$ curves for good and poor quality samples (Coutinho; Oliveira; Oliveira, 1998).

Soil disturbance affects the compression curve of the consolidation test, as shown in Figure 2.20 for Recife and Rio de Janeiro clays. A poor quality sample, when compared with a good quality sample, will present lower overconsolidation stress. The change of the void ratio, related to a change of the effective stress is also altered by disturbance. This can lead to predicted settlements being different from the actual settlements and wrong fill volume values in projects for temporary surcharge and for the compensation of these settlements.

It is also noted that the void ratio versus the logarithm of the effective stress becomes linear with disturbance, but for high effective stress values the compression curves are similar. The disturbance of a sample decreases the permeability and consequently the coefficient of vertical consolidation, which can cause an incorrect evaluation of settlement rate, i.e., the expected stabilization period based on disturbed samples can be greater.

## 2.8 TRIAXIAL TESTS

The strength and modulus values measured in UU triaxial tests are influenced by the stress relief process and disturbance. However, considering the relatively small cost, they provide additional data to obtain the $S_u$ profile.

The isotropic triaxial consolidation test (CIU) is not widely used in Brazilian practice. In some special projects anisotropic consolidation triaxial tests (CAU) are carried out. In this case, the effective vertical $\sigma'_{vo}$ and horizontal $\sigma'_{h0}$ stresses *in situ* for which the specimen will be consolidated must be previously estimated (see Figure 2.17). Considering that $\sigma'_{h0} = K_0 \cdot \sigma'_{vo}$ and the earth pressure coefficient at rest $K_0$ can be defined by the equation $K_0 = (1 - \sin \phi')$. $OCR^{\sin \phi'}$ for the CAU test, both OCR and friction angle $\phi'$ of the soil should be previously known or estimated. The CAU tests require more time, unusual equipment and procedures, and are often performed by specialized laboratories.

## 2.9 FINAL REMARKS

The Geotechnical site investigation in clusters allow the integrated analysis of all in situ and laboratory test results, thus a better overall understanding of the geomechanical behavior of soft soil deposit, besides the assessment of the consistency of different test results, as shown in Figure 2.21. In this specific case (Figure 2.21), the values for overconsolidation stresses being lower than the *in situ* stresses indicate sample disturbance, for example.

Figure 2.21 shows the results of the tests types described herein, *in situ* (SPT, vane and piezocone) and laboratory (index tests, oedometer tests, triaxial UU tests). Regarding the stratigraphy, SPT and piezocone tests clearly complement each other. The same can be said about the design strength profile $S_u$, in this case, combining data from vane, piezocone and UU tests.

The coefficient of consolidation values obtained from oedometer and piezocone tests also complement each other, but oedometer tests are fundamental because they provides compressibility parameter. For the estimation of design values of coefficients of consolidation, the piezocone test is recommended as well as the consolidation test.

Figure 2.22 presents consolidation coefficients estimated from piezocone and oedometer tests and monitoring of soft soil deposits in Western areas of Rio de Janeiro. This great variability of $c_v$ values observed is not uncommon for clays in the Southeast of Brazil.

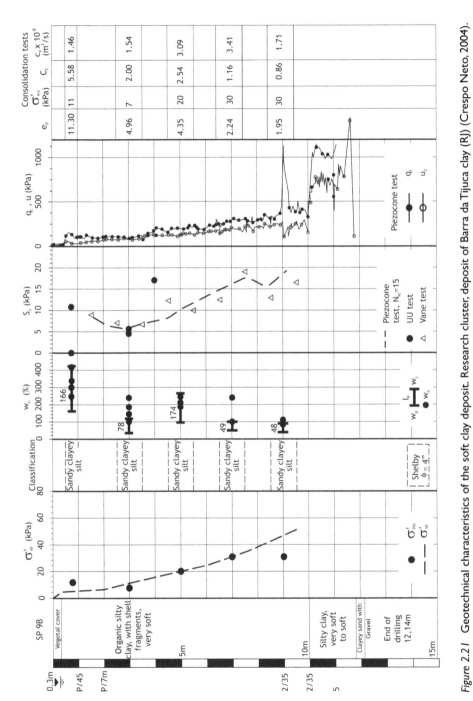

Figure 2.21 Geotechnical characteristics of the soft clay deposit. Research cluster, deposit of Barra da Tijuca clay (RJ) (Crespo Neto, 2004).

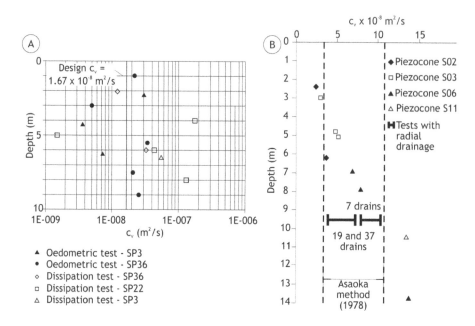

*Figure 2.22* Consolidation coefficient profile from piezocone and oedometer tests at the western area of Rio de Janeiro – normally consolidated range: (A) Recreio; (B) Barra da Tijuca (Almeida et al., 2001).

Geotechnical parameters of some Brazilian deposits, which are useful for preliminary project calculations, are presented in the Appendix. Geotechnical parameters for the Baixada Santista (Santo Plains) deposits are discussed in detail by Massad (2009).

# Chapter 3

# Geotechnical properties of very soft soils: Rio de Janeiro soft clays

This chapter describes the geotechnical properties of very soft clays illustrated by the case Sarapuí clay in Rio de Janeiro, probably the best studied soft clay deposit in Brazil (Almeida and Marques, 2003). The overall elasto-plastic behavior of soft clays is previously presented for background reference.

## 3.1 OVERALL BEHAVIOR OF VERY SOFT SOILS: CAM CLAY MODELS

Overall behavior of very soft clays may be consistently described by the elasto-plastic Cam-clay models developed as part of the "Critical State Theory" (Roscoe et al., 1958) by Cambridge University researchers in the sixties (Schofield and Wroth, 1968; Roscoe and Burland, 1968). Since then, further developments of the Cam-clay models have been proposed (e.g., Koskinen et al., 2002; Wheeler et al., 2003) while maintaining the principles of the original models. For lightly overconsolidated non-structured very soft clays the original models have the benefits of the simplicity and representative performance.

Critical state conditions may be observed in triaxial tests at large strain conditions at which there is no change in void ratio and effective stresses. Critical state can be directly and easily reached in undrained or drained tests, on normally consolidated or lightly over-consolidated samples, or even in highly over-consolidated clays if large strains can be reached in homogeneous conditions. The latter are of little interest for the scope of this book.

### 3.1.1 Stress and strain variables

The three variables used in Cam-clay models (simplified for triaxial test conditions) are:

the mean effective stress

$$p' = \frac{(\sigma'_a + 2\sigma'_r)}{3} \tag{3.1}$$

the deviator stress

$$q = \sigma'_a - \sigma'_r \tag{3.2}$$

and the specific volume

$$v = 1 + e \tag{3.3}$$

where $\sigma'_a$ and $\sigma'_r$ are, respectively, axial and radial effective stresses corresponding to principal effective stresses in axi-symmetric conditions, v is the specific volume and e is the voids ratio.

It is more convenient to represent the stress variables p′ and q in terms of $\sigma'_a$ and $\sigma'_r$ than in terms of principal stresses as it facilitates to differentiate between compression and extension paths, respectively, positive and negative using the definition of q presented in (3.2). For numerical analyses p′ and q may be written in three-dimensional conditions (Atkinson and Bransby, 1978; Wood, 1990).

Strains used in Cam-clay models are essentially shear strain $\varepsilon_s$ and volumetric strain $\varepsilon_v$ that, for overall consistency in terms of stress and strain invariants, are defined as (Atkinson and Bransby, 1978)

$$\varepsilon_s = \frac{2}{3}(\varepsilon_a - \varepsilon_r) \tag{3.4}$$

$$\varepsilon_v = \varepsilon_a + 2\varepsilon_r \tag{3.5}$$

where $\varepsilon_a$ and $\varepsilon_r$ are respectively, axial and radial strains measured in the triaxial test.

### 3.1.2  Model parameters

Cam-clay models incorporate five soil parameters obtained from standard laboratory tests, mainly triaxial, and isotropic or oedometer compression tests.

The two compression parameters used in Cam-clay models are the compression index $\lambda$ and the recompression or swelling index $\kappa$ shown in the e versus ln p′ plot of isotropic compression ($\sigma'_1 = \sigma'_2 = \sigma'_3$) in Figure 3.1. These can be related to the compression index $C_c$ and the recompression index $C_s$ in the e versus $\log \sigma'_v$ plot by

$$\lambda = \frac{C_c}{2.3} \tag{3.6}$$

$$\kappa = \frac{C_s}{2.3} \tag{3.7}$$

The critical state line is formally parallel to the isotropic compression line and it is represented in Figure 3.1.

The third parameter is $\Gamma$, the specific volume corresponding to p′ = 1 kPa which is necessary to locate the critical state line in the v versus p′ plot.

Figure 3.1 also shows the parameter N, which is the specific volume corresponding to p′ = 1 kPa necessary to locate the isotropic compression line in the v versus p′ plot. However, N is not an independent variable as it depends on the actual Cam-clay model chosen. For the more widely used modified Cam-clay model, N is defined as

$$N = \Gamma - \ln 2(\lambda - \kappa) \tag{3.8}$$

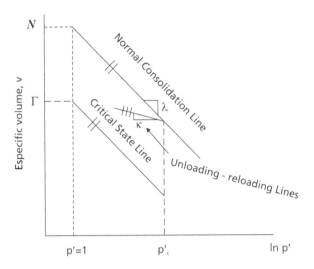

Figure 3.1 Normal Consolidation and critical state lines in the $v - \ln p'$ plane, isotropic compression.

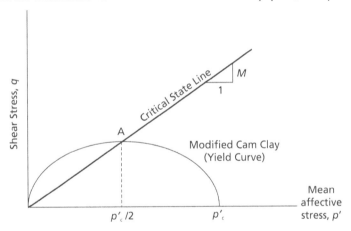

Figure 3.2 Yield surface of a Cam clay model in the $q-p'$ plane.

The parameter M shown in Figure 3.2 defines the inclination of the critical state line in the p'–q plot and may be defined as

$$M = 6 \frac{\sin \phi'_{cs}}{(3 - \sin \phi'_{cs})} \tag{3.9a}$$

where $\phi'_{cs}$ is the friction angle corresponding to critical state (constant volume) conditions.

In practical terms it is possible to adopt the simplified empirical equation

$$M = \frac{\phi'_{cs}}{25} \tag{3.9b}$$

where values of $\phi'_{cs}$ are in degrees.

The yield curve separates elastic conditions from elasto-plastic conditions. For the modified Cam-clay model the yield curve is elliptical as shown in Figure 3.2 and is defined by

$$\frac{q}{p'^2} + M^2 \cdot \left(1 - \frac{p'_c}{q'}\right) = 0 \tag{3.10}$$

where $p'_c$ is the preconsolidation or yield stress at the isotropic compression line (see also Figure 3.1).

The fifth Cam-clay parameter is the shear modulus G defined as

$$G = \frac{E}{2(1 + v)} \tag{3.11}$$

where E is the Young's modulus and $v$ is the Poisson's ratio. As water does not transmit shear, thus the shear modulus is equal for total and effective stress conditions, i.e., $G = G'$. In other words, G may be computed from drained or undrained tests and is just simpler and quicker to obtain G from undrained triaxial tests. Replacing in (3.11) the value of Poisson's ratio for undrained condition $v_u = 0.5$,

$$G = \frac{E_u}{3} \tag{3.12}$$

And the undrained modulus $E_u$ may be obtained from consolidated undrained CU triaxial tests.

Considering the large strains and low factors of safety commonly associated with construction in very soft clays, the value of $E_u$ adopted to compute G adopted is usually the secant modulus $E_{u50}$ corresponding to the 50% of the maximum deviator stress q.

The bulk effective modulus K' relates increments of elastic volumetric strains with the increment of volumetric stresses p'. In Cam-clay models the equation of bulk modulus K may be obtained from the unloading-reloading line $v = v_\kappa - \kappa \ln p'$ indicated in Figure 3.1.

$$K' = \frac{(1 + e)\, p'}{\kappa} \tag{3.13}$$

Thus K' depends on the mean effective stress p', void ratio e and the slope $\kappa$ of the unloading line.

The shear modulus G may be also defined (Atkinson and Bransby, 1978) as

$$G = \frac{3K'(1 - 2v')}{2(1 + v')} \tag{3.14}$$

where $v'$ is the Poisson's ratio in terms of effective stresses.

Substituting (3.13) into (3.14) one gets

$$G = \frac{3(1 - 2v')(1 + e)p'}{2(1 + v')\kappa} \tag{3.15}$$

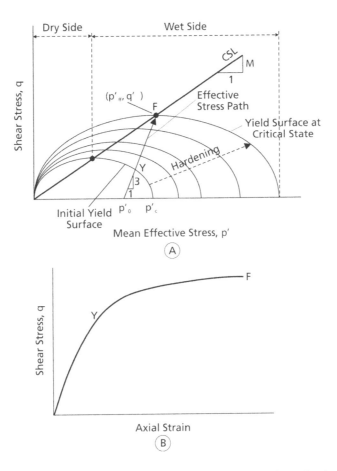

*Figure 3.3* Cam clay hardening behavior: (A) evolution of a yield surface during hardening; (B) stress–strain curve with strain hardening.

Therefore (3.15) considers the modulus G as stress dependent which may be more consistent than assuming a constant G as in (3.12). Under elastic conditions, which can be associated to the unloading-reloading stages, a constant Poisson's ratio $\nu'$ is usually assumed in (3.15) for simplicity.

### 3.1.3  Yield conditions

For the more usual loading conditions related to very soft clays the yield curve increases as loading progresses and this is exemplified in the diagram shown in Figure 3.3. This enlargement of the yield curve is the strain-hardening behavior shown in the $p'$–$q$ plot of Figure 3.3A which shows also the stress path of a lightly overconsolidated specimen (yield stress $p'_c$) for conventional triaxial drained shearing ($\Delta q / \Delta p' = 3$) starting at $p'_a$, the first yielding occurring at point Y and experiencing continuous hardening to reach critical state condition at point F. Figure 3.3B also shows the hardening behavior in

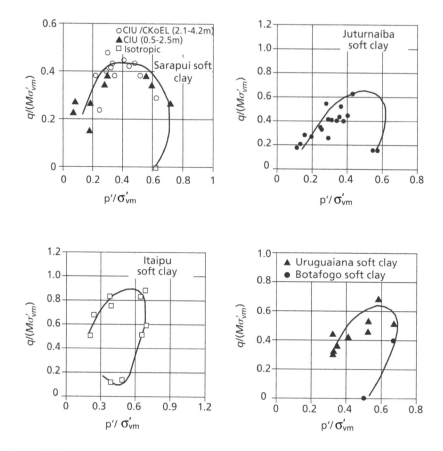

*Figure 3.4* Limit state curves of five Rio de Janeiro soft clays (Futai, 1999; Futai et al. 2008).

terms of the stress strain curve. The yielding point Y and the large strains critical state conditions are also shown. The strain-softening behavior is the decrease of the yield curve and q following post peak. This is usually related to overconsolidated conditions less commonly observed in very soft clays.

The original Cam-clay models were proposed for isotropic yield conditions. This means a yield curve symmetric with respect to the isotropic line p', as shown in Figure 3.2. However, natural clays, contrary to laboratory prepared clays are anisotropic, have different horizontal and vertical properties, due to their formation process, thus yield curves of natural clays are not centered on isotropic axis (e.g., Diaz-Rodriguez et al., 1992; Koskinen et al., 2002; Wheeler et al., 2003). Figure 3.4 shows normalized yield curves of five natural soft clays localized at Rio de Janeiro State (Futai, 1999).

Both q and p' were normalized with relation to preconsolidation pressure, however q was also normalized with relation to the critical state parameter M, defined in Eq. 3.9, and varying with $\phi'$ (friction angle) as function of the tests depth.

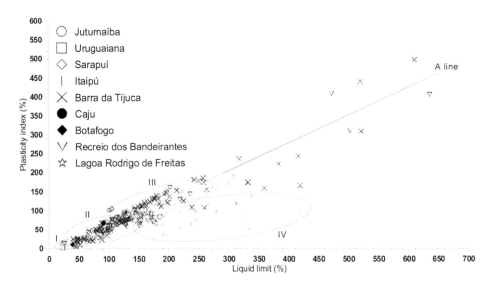

*Figure 3.5* Plasticity chart for Rio de Janeiro clays (Mello, 2013; adapted from Futai, 1999).

## 3.2   INDEX PROPERTIES OF SOME RIO DE JANEIRO CLAYS

Index properties of Rio de Janeiro clays (Almeida et al., 2008) are summarized here. Geotechnical properties and characteristics of soft clay deposits in the west region of Rio de Janeiro (Almeida et al., 2010) are presented in Table A.2 (Annex). It may be observed the high values of the compression ratio CR observed in most sites.

The relationship between the $I_P$ and the liquid limit $w_L$ has been traditionally used to classify fine-grained soils. The usual functional relationship between $I_P$ and $w_L$ is:

$$I_P = A(w_L - B) \tag{3.16}$$

The Casagrande A line for soil classification gives $A = 0.73$ and $B = 20$, as shown in Figure 3.5. For a large set of data of 520 soils and for $I_P$ and $w_L$ varying between 10–90% and 25–120%, respectively, Nagaraj and Jayadeva (1983) obtained $A = 0.74$ and $B = 8$ for organic soils, yielding a line slightly above the Casagrande line, which still agrees fairly well with data of Rio de Janeiro clays for soils with lower water content.

Four regions in the Plasticity chart were proposed by Futai (1999) and are represented in Figure 3.5 and described in Table 3.1. Three of them are rather well fitted by line A, while region IV is well outside the range of equation 3.1, for Itaipú clay and also for Juturnaíba clay, due to their high value of organic matter. This classification is also a function of compressibility and the range of the strength parameters, as shown in Table 3.1.

Some very soft clays exhibit high organic content and Figure 3.6 shows a relationship between water content and organic matter for Rio de Janeiro clays. It is well known (e.g., Schofield and Wroth, 1968) that the water content is directly related to compressibility and undrained strength. Therefore, organic content influences directly compressibility and strength.

Mitchell (1993) has shown the influence of the organic content in increasing the Atterberg limits and decreasing both the dry density and the undrained strength. For Sarapuí clay, Coutinho and Lacerda (1987) have clearly shown the influence of the organic content in increasing compression index $C_c$, compression ratio $CR = C_c / (1 + e_o)$, and secondary compression index $C_{\alpha\varepsilon} = C_\alpha / (1 + e_o)$.

*Table 3.1* Classification of Rio de Janeiro clays (Futai, 1999).

| Classification | Consistency | Strength | $e_o$ | $I_P$ | $w_L$ (%) | $C_c/(1 + e_o)$ | $S_u$ (kPa) | $\phi'$ (°) | Rio de Janeiro clays |
|---|---|---|---|---|---|---|---|---|---|
| Region I | Stiff inorganic clays | High undrained strength | <2 | <10 | <40 | 0.15–0.35 | >50 | 28–40 | Botafogo Uruguaiana |
| Region II | Slightly soft organic clays | Low undrained strength | 2–4 | 10–120 | 30–200 | 0.25–0.35 | 6–15 | 25–35 | Cajú, Barra da Tijuca, Sarapuí |
| Region III | Medium soft organic clays | Low undrained strength | 4–6 | >80 | >100 | 0.40–0.60 | 6–25 | 30–40 | Juturnaíba, Sarapuí |
| Region IV | Very soft organic clays – peat | Low undrained strength | >3.5 | >130 | >150 | 0.25–0.35 | 10–25 | <65 | Itaipú |

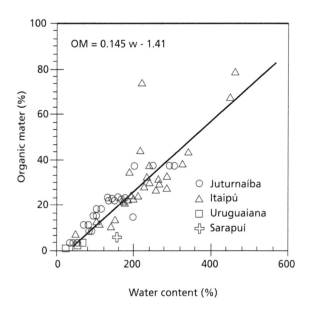

*Figure 3.6* Organic matter content and water content relationship for Rio de Janeiro clays (Futai, 1999).

## 3.3  COMPRESSIBILITY AND STRESS HISTORY

### 3.3.1  Compressibility

Sarapuí clay is a very compressible soil, for which $C_c$ varies from 1.3 to 3.2 and average $C_s/C_c$ is 0.12 as shown in Figure 3.7. The average $C_c/(1 + e_0)$ relationship is 0.41, thus an indication that this clay is bound to present relevant viscous behaviour.

As soft clays in general, and particularly for structured clays the behaviour of Sarapuí clay under uni-dimensional consolidation is strongly affected by sampling disturbance. This effect is shown in Figure 3.7 for tests performed on good and bad quality samples as well as on laboratory remolded samples of Sarapuí clay (Ferreira & Coutinho, 1988; Coutinho et al., 1998).

Coutinho (1976) carried out conventional and special oedometer tests with radial drainage and measured coefficient of horizontal consolidation ($c_h$) of samples at normally consolidated range, and observed decrease of $c_h$ with sample disturbance, and

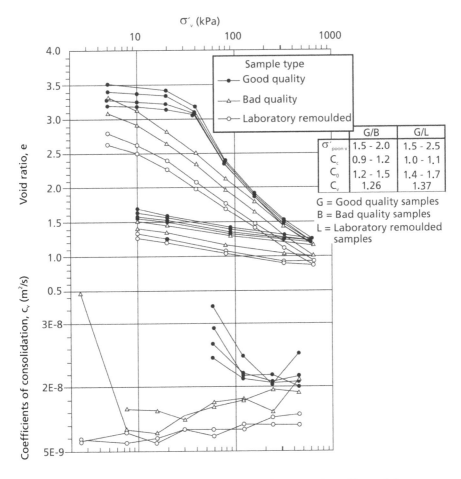

*Figure 3.7* Sampling effect on uni-dimensional compression of Sarapuí clay.

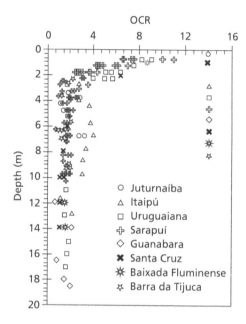

*Figure 3.8* OCR profiles of Rio de Janeiro clays (Futai, 1999).

obtained $c_{h\,undisturbed}/c_{h\,remolded} = 1.42$ (at 5.8 m) and 1.55 (at 6.8 m), which data are quite relevant with respect to smear effects due to vertical drains installation.

The $(e_{vo} - e_o)/e_o$ relationship, where $e_{vo}$ is the void ratio at *in situ* vertical stress, is an indication of sample quality. Lunne et al. (1997) proposed that for high quality samples this relationship is lower then 0.04, for an OCR between 1 and 2. Analysis of 63 oedometer tests carried out by Ortigão (1980) shows an average $(e_{vo} - e_o)/e_o$ relationship of 0.033 and for only 16% of samples this relationship is higher than 0.04.

### 3.3.2  Overconsolidation ratio (OCR)

The variation of the overconsolidation ratio OCR with depth, for eight Rio de Janeiro soft clay deposits, is shown in Figure 3.8. The OCR profile for all clays is within a narrow range, which suggests that the stress histories of the clay deposits in the Rio de Janeiro region are similar. It is observed that the top 4 m present larger OCR variation but below 4 m the OCR is in average ranging from 1.5 to 2 and mostly constant which is mainly due to secondary compression (ageing), while the continuous decrease of OCR with depth is attributed to changes in ground water level (Parry and Wroth, 1981).

### 3.4  HYDRAULIC CONDUCTIVITY AND COEFFICIENT OF CONSOLIDATION

Conventional and special oedometer tests were carried out on Sarapuí clay with radial drainage: inflow and outflow types, and inflow-outflow tests also described as double

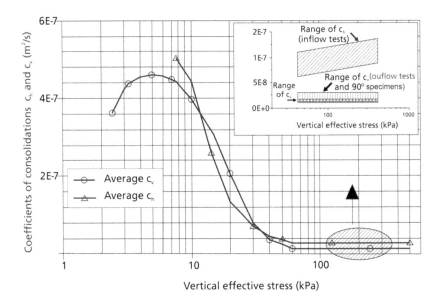

*Figure 3.9* $c_h$ and $c_v$ values from laboratory tests at 5.5–6.0 and 6.5–7.0 m depth Sarapuí clay (Coutinho, 1976).

drainage, and with vertical drainage on samples at 90° with horizontal plane, in order to obtain coefficient of horizontal consolidation ($c_h$) value (Coutinho, 1976; Lacerda et al., 1977, 1995). For tests carried out with radial drainage with inflow and outflow, $c_h$ values were computed using Barron (1948) and Scott (1961) solution, respectively, both with "equal strain" consideration. Figure 3.9 shows average vertical and horizontal coefficients of consolidation with stress, for those tests, carried out on undisturbed samples collected at the middle of the clay layer.

For higher stress levels, showed in detail in Figure 3.9, the $c_h$ range of inflow tests are systematically higher than outflow tests, maybe due to leaks or formation of smear zone around the external boundary of the sample, as vertical deformation develops. The outcome is that outflow tests and double drainage tests are quicker than inflow tests and are not affected by possible flow between top plate and the ring (Lacerda et al., 1995). The $c_h/c_v$ relationship lies between 1.0 and 2.0 at normally consolidated range, which is typical for very soft clays.

Hydraulic conductivity at overconsolidated domain, $k_v$, at the middle of the clay deposit lies in the range 2 to $5 \times 10^{-9}$ m/s (Coutinho, 1976), which is much smaller than measured (Gerscovich et al., 1986) at the crust. On the overconsolidated domain, $k_v$ values measured at constant head tests carried out on triaxial cells lie in the range of $3 \times 10^{-9}$ m/s (at 1.5 m) to $24 \times 10^{-8}$ m/s.

Figure 3.10 shows permeability variation with void index of Sarapuí clay. At normally consolidated domain anisotropy permeability factor was $k_h/k_v = 1.6$ to 2.1 for void index between 3 and 2. Sandroni et al. (1997) found $k_{h0}/k_{v0}$ close to 2 for a northwestern Brazilian organic clay. However, for inorganic eastern Canadian clays, for example, this factor is lower, of about 1.1 (Leroueil et al., 1983).

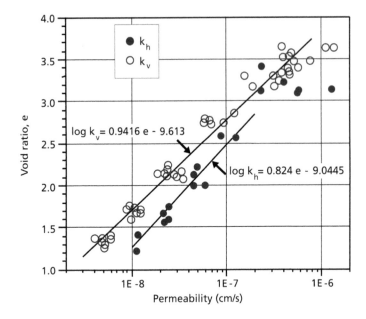

*Figure 3.10* Vertical and horizontal permeability variation with void index – Sarapuí clay (Lacerda et al., 1977).

## 3.5   SOIL STRENGTH

### 3.5.1   Undrained strength – laboratory and in situ data

Figure 3.11A shows the average undrained strength ($S_u$) profile of Sarapuí clay obtained from UU, $CK_0U$ and CIU (Shansep method) tests performed in the early eighties but yet quite useful to understand the overall behavior a typical Brazilian soft clay. For tests performed using the Shansep method, the samples were brought to consolidation until normally consolidated domain was reached, and then unloaded to an OCR value, and then undrained compression was carried out until failure. It was observed that $S_u$ values obtained using Shansep method are underestimated, since this method may induce the destruction of the natural clay structure, apart from the fact that samples may be partly disturbed. $S_u$ from triaxial tests are lower when compared to vane tests values.

The critical state strength profile (Figure 3.11A) was obtained using critical state parameters and OCR and $K_0$ values varying with depth (Almeida, 1982). $S_u$ values predicted from isotropic consolidation are higher than those from anisotropic consolidation, the same behaviour observed when using Shansep method.

The undrained strength obtained from UU tests performed on samples of the crust lies between 4 and 7 kPa (Bressani, 1983; Gerscovich, 1983). These are lower than those measured by vane tests shown in Figure 3.11B. It is possible that due to high permeability of the crust, strength measured at vane test at this level was partially drained.

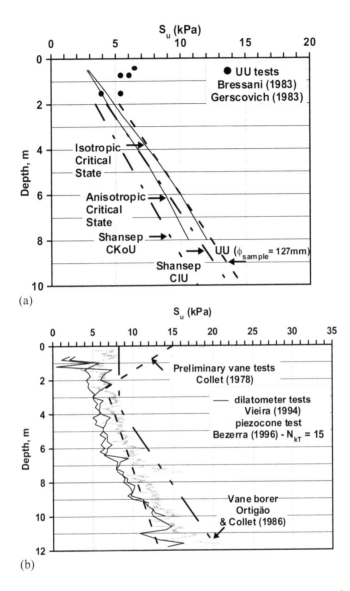

*Figure 3.11* Undrained strength profiles of Sarapuí clay: (a) laboratory tests and stress history equations – CIU and CK$_0$U tests (Ortigão, 1980); (b) in situ tests profiles.

Natural clays are anisotropic and strength parameters can change with direction. In order to study anisotropy vane tests using different sizes of vane blade (H/D = 2, 0.5 and 4) were performed on Sarapuí clay (Collet, 1978; Costa Filho et al., 1977). From these simplified studies, where limitations are recognized nowadays (Wroth, 1984), it seemed that strength anisotropy of Sarapuí clay is not important.

Collet (1978) carried out a series of vane tests at the Sarapuí site, and obtained the S$_u$ average profile shown in Figure 3.11B. Ortigão & Collet (1986) improved the

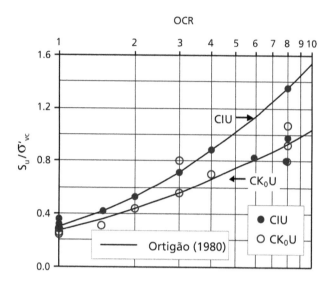

*Figure 3.12* Normalized undrained versus OCR – CIU and $CK_0U$ tests (Ortigão, 1980).

vane test equipment to decrease friction and obtained another $S_u$ average profile, with $S_u$ values higher than before (Figure 3.11B). Using the new vane data, the normalized $S_u/\sigma'_{vm}$ average relationship is 0.35 (Almeida, 1986), in accordance with the relationship proposed by Leroueil et al. (1983), $S_u/\sigma'_{vm} = 0.2 + 0.0024I_P$, for eastern Canadian clays, where $\sigma'_{vm}$ is the vertical preconsolidation stress.

Values of $S_u$ deduced from dilatometer tests (Vieira 1994) are lower than $S_u$ values from vane tests. Differences from *in situ* tests are probably due to different stress paths and empirical factors of piezocone and dilatometer. It is well known (Wroth, 1984) that each *in situ* or laboratory test follows a certain stress path, which then results in a given undrained strength $S_u$.

Normalized $S_u$ increases with OCR, in a different way for CIU and $CK_0U$ tests (Figure 3.12). For OCR = 1, the curves present the same $S_u/\sigma'_{vc}$, but as OCR increases, for $CK_0U$ tests $S_u/\sigma'_{vc}$ relationship is lower. The variation of the normalized strength with OCR has been well predicted using critical state parameters ($\lambda$, $\kappa$, M) for both isotropic and anisotropic consolidation conditions (Almeida, 1982). As expected, and seen in Figures 3.11A and 3.12, for triaxial tests carried out using Shansep method, CIU tests presented systematically higher $S_u$ normalized value than $CK_0U$ tests.

## 3.5.2 Effective strength parameters

Figure 3.13 shows the stress path of CIU tests carried out at different consolidation pressure ($\sigma'_{vc}$) (Ortigão, 1980); stress state at rupture from $CK_0U$ tests (Ortigão, 1980) and CIU tests (Costa Filho et al., 1977), all performed on samples below crust. The average strength envelope was obtained from $t_{max} = ((\sigma'_1 - \sigma'_3)/2)_{max}$ values. The friction angle of Sarapuí clay is $\phi' = 25°$ and effective cohesion varies between $c' = 0$ and $c' = 1.5$ kPa, as shown in Figure 3.13 (Costa Filho et al., 1985; Gerscovich et al., 1986).

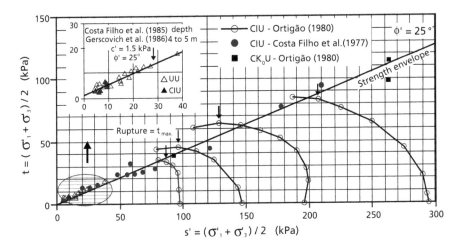

*Figure 3.13* Strength envelopes of Sarapuí soft clay.

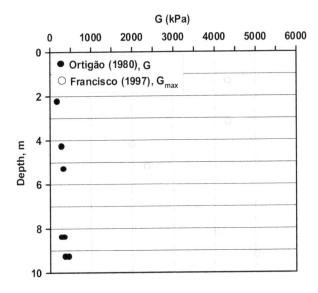

*Figure 3.14* Variation of G and $G_{max}$ with depth.

The strength envelope obtained (Gerscovich, 1983; Bressani, 1983) for the dessi-cated crust soil (0.2 m to 1.0 m) resulted in strength parameters $c' = 0$ and $\phi' = 31°$, thus higher than that of the clay layer below.

## 3.6 DEFORMATION DATA

Francisco (1997) performed seismic cone penetration tests at Sarapuí site using a seismic CPT with geophones at cone tip in order to obtain maximum shear stiffness

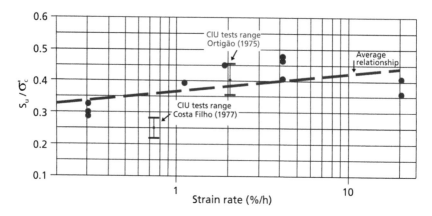

*Figure 3.15* Variation of normalized undrained strength of Sarapuí clay with strain-rate (Ortigão, 1980).

($G_{max}$) variation with depth. These results compared with results from UU triaxial tests, $G = E_u/3$ (Ortigão 1980) show, as expected (see Figure 3.14), quite low G values obtained from UU tests. A $G/G_{max}$ profile was obtained (Francisco, 1997), and thus G values can be inferred from $G_{max}$ values.

## 3.7   VISCOUS BEHAVIOR

### 3.7.1   Strain rate during shearing

Undrained shear strength changes from 5 to 20% per logarithm cycle of strain rate, and a typical increase of 10% per logarithm cycle of strain rate (Leroueil & Marques, 1996) is usually assumed. Evidence of strain rate effect on Sarapuí clay normalized $S_u$ (with confining stress $\sigma'_c$) was presented by Ortigão (1980), and an average increase of 15% per logarithm cycle of strain rate was measured on CIU tests performed in normally consolidated samples, as shown in Figure 3.15.

### 3.7.2   Strain rate during constant loading oedometer tests

Feijó (1991) carried out long-term oedometer tests, under controlled temperature, in order to study the secondary expansion of samples under different OCR values. Figure 3.16 shows volumetric strains variation with time. First, the samples were loaded in the overconsolidated range and then unloaded to different values of OCR. For OCR between 1.5 and 2 after some expansion, the sample begins to compress again. For OCR of 8 and 12 there is secondary expansion during the test.

For OCR between 2 and 6, secondary swelling is not significant and the sample is supposed to be at an equilibrated state, where strain rate is almost zero (Figure 3.17A). Under this OCR range, the coefficient of earth pressure lies between 0.77 and 1.23, which is the range of $K_0$ at equilibrium state (Figure 3.17B).

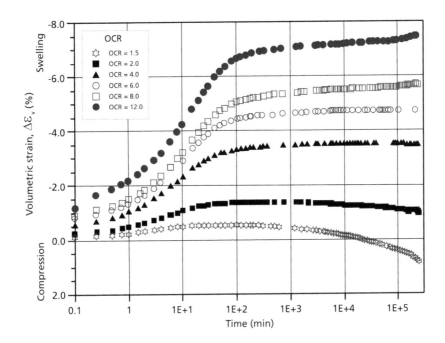

*Figure 3.16* Volumetric strains observed after unloading of Sarapuí clay (Feijó, 1991).

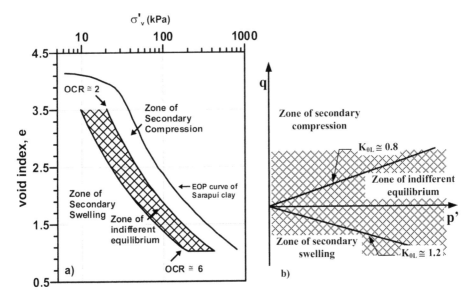

*Figure 3.17* Zone of indifferent equilibrium of Sarapuí clay (Feijó & Martins, 1993).

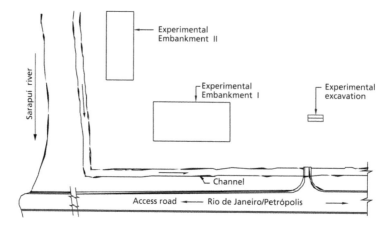

*Figure 3.18* Sarapuí test site.

This behavior is quite important for design of pre-loading of embankments, since it is possible to decrease secondary compression of clay, when unloading from this OCR range.

## 3.8   FIELD STUDIES

Two test embankments were built in the 70' as shown in Figure 3.18. The Embankment I was built until rupture, 30 days after the beginning of loading in December/1977. Embankment II was built in November/1980 and heightened five years later and monitored during almost 10 years after first loading. The majority of the studies described above were associated with the design of these trials, and an extensive comparative analysis between laboratory and *in situ* test results and field results was performed (Ortigão et al., 1983; Almeida et al., 1989).

### 3.8.1   Embankment I

Three sections of the embankment were instrumented with hydraulic piezometers, inclinometers, horizontal extensometers, settlement plates and surface marks (Ortigão et al., 1983).

The beginning of fissuring occurred when embankment was at the height of 2.5 m. At this time the strain rate increased rapidly and at a height of 2.8 m fissures were 5 cm wide and at 3.1 m the rupture was generalized. Subsequent analysis (Almeida, 1985; Sandroni, 1993) of this case history indicated that the actual failure took place when the embankment was the height of 2.5 m. It is well known that for design purposes a Bjerrum type correction has to be applied to the $S_u$ value measured in vane tests. The re-analysis of Embankment I failure using a 3D failure surface has shown (Sandroni, 1993; Pinto, 1992) that the correction factor for Sarapuí clay is $\mu = 0.70$, which lies slightly above (for a typical $I_P = 80$) the $\mu$ versus $I_P$ relationship proposed by Azzouz et al. (1983), for this type of failure.

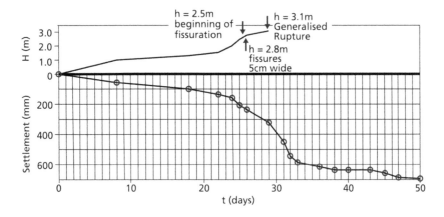

*Figure 3.19* Settlement curve – Embankment I.

Figure 3.19 shows the variation of settlement and height of Embankment I with time, for settlement plates at the centerline of the embankment. At the beginning of fissures (H = 2.5 m) the embankment was not maintained at this height for more than one day, therefore there was no time for deformations due to this loading to develop. The failure could have occurred at this height, if it was maintained for a longer period, before increasing to 2.8 m. The rupture developed slowly, in a progressive way, so it may have mobilized larger strains, at smaller $S_u$ values, going to critical state.

During construction time (30 days) the clay deposit was partially drained, particularly the crust (Ortigão et al., 1983; Gerscovich et al., 1986). This consideration implies in significant increase of the crust's $S_u$, when back-analysing the clay failure.

The hypothesis of partial drainage is consistent with piezometric measures at the middle of the embankment and at the top of the deposit. There was pore-pressure dissipation during early stages of construction.

Embankment I was analysed numerically (Almeida, 1981; Almeida & Ortigão, 1982) using the modified Cam-clay model. As shown in Figure 3.4, the yield surface of Sarapuí clay is not far from isotropic. Moreover, this model has provided good insight for other case histories in Brazil (Antunes Filho, 1996) and was therefore also used here. Analyses were performed for both, undrained and partially drained conditions (using Biot 2D consolidation theory) using the same critical state parameters, but with vertical and horizontal hydraulic conductivities for the partially drained case. Good overall agreement between measured and numerical values of base settlements and pore pressures was observed just for the partially drained case (coupled consolidation), but not for the undrained condition, as shown in Figure 3.20, which suggests that even for embankment loading during one month, some drainage may have taken place.

## 3.8.2  Embankment II

Embankment II was built 35 m wide and 315 m long, divided in seven-instrumented sub-areas, where different types of drains were installed, as described in Table 3.1.

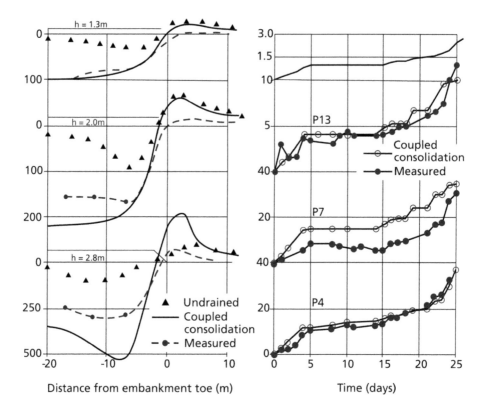

*Figure 3.20* Settlements at the embankment I base and pore pressure variation, for stages of construction.

Loading was applied in two main stages and the final height was about $h = 3.6$ m. The first stage of loading was applied during 1981, in two steps. The first step ($h = 0.7$ m) during a month and the second, 200 days after, lasted approximately 2.5 months ($h = 2.0$ m). The second stage, in 1986, was applied only from section B to G, in one week, until final height.

Pore pressure dissipation showed that top layer had a higher coefficient of consolidation, as expected from tests on this layer (Gerscovich et al., 1986).

For the first stage of loading, when range of vertical effective stress was going from 15 to 60 kPa, the $c_v$ from field data was difficult to analyze and compare. At this stress range there is an important variation of $c_v$ when soil goes from overconsolidated state to normally consolidated state, as shown in oedometer results presented on Figure 3.9.

For the second stage of loading, from 60 to 80 kPa, at normally consolidation range, the $c_v$ from oedometer tests remained almost constant with vertical stress (Figure 3.9).

Table 3.2 shows average coefficient of vertical and horizontal consolidation obtained from laboratory, *in situ* tests and back-calculated from settlement data using

Table 3.2   $C_h$ and $c_v$ data (adapted from Almeida et al., 1992).

| Test or data | Depth (m) | Method | Reference | $c_v$ $10^{-8}\,m^2/s$ | $c_h$ $10^{-8}\,m^2/s$ |
|---|---|---|---|---|---|
| Oedometer test | 5–6 | Taylor | Coutinho (1976) | 1.2 | 2.4 |
| Piezocone | 2.2–8.2 | Houlsby & Teh (1988) | Danziger (1990) | 1.6–4.4 | 3.1–8.7 |
| Field instrumentation: settlement plates | whole layer | Asaoka (1978) | Schmidt (1992) | 17.8* | 3.1–4.4** |
| Field instrumentation: top settlement magnetic gages | whole layer | Asaoka (1978) | Almeida et al. (1989) | 22.6 | 4.2–8.1*** |
| Field instrumentation: Casagrande piezometers | 3.3–8.3 | Orleach (1983) | Ferreira (1991) | 2.2–4.5 | 1.2–2.8 |

*average value of central settlement plates.
**average value of central settlement plates (no smear, vertical drainage considered).
***smear but no vertical drainage considered.

Asaoka (1978) method, for the second stage of loading. From those results it was observed that $c_{v(field)}/c_{v(laboratory)}$ relationship range lies between 20 and 30.

Schmidt (1992) showed that low value of $\sigma'_{vf}/\sigma'_{v0}$ can lead to error on $c_v$ determination with Asaoka's method, as secondary consolidation gets quite important. The good agreement between $c_v$ values obtained from laboratory tests, *in situ* tests, and settlement site plates data by (Almeida et al., 1993a) for another site close to Sarapuí was due to the high $\Delta\sigma'_v$ (24 m height fill), since almost no secondary compression was observed.

Pinto (2001) discussed the validity of Asaoka's method and observed that $c_v$ values, as well as final settlement value, computed using the method are very susceptible to monitoring time. From series of data analyzed within different periods of observation, for the series with 100 days of observation the $c_v$ computed was 0.082 m²/day, while after 4050 days, the $c_v$ was 0.005 m²/day.

Chapter 4

# Prediction of settlements and horizontal displacements

This chapter deals with the calculation of settlements and their variation over time (without PVDs) and the prediction of horizontal displacements.

## 4.1 TYPES OF SETTLEMENTS

Settlements are usually divided into immediate settlements ($\Delta h_i$), primary consolidation settlements ($\Delta h$) and secondary compression settlements ($\Delta h_{sec}$), as schematically presented in Figure 4.1.

This classification of settlements is convenient for calculations, but it may be considered simplistic. Alternatively, settlements may be classified as construction settlements and long term settlements (Leroueil, 1994). Construction settlements are the sum of immediate settlements, $\Delta h_i$, and of primary recompression settlements, $\Delta h_{rec}$ (from in situ stress to overconsolidation stress). Long-term settlements are the sum of

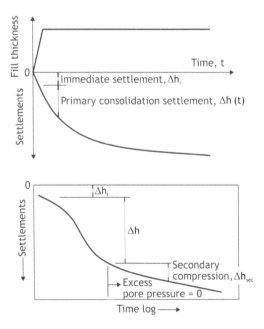

Figure 4.1 Types of settlements (Rixner, Kraemer and Smith, 1986).

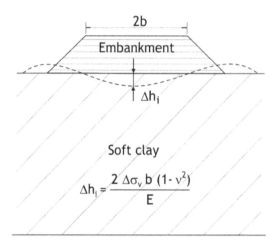

*Figure 4.2* Immediate consolidation settlement; scheme of vertical displacements on the base of the embankment (Poulos and Davis, 1974).

settlements by virgin primary consolidation, $\Delta h_{vc}$ and secondary compression settlements, $\Delta h_{sec}$. This classification is more realistic than the previous one, because, on one hand, the immediate settlement, $\Delta h_i$, is combined with the recompression consolidation settlements (line $C_s$), associated with greater values of consolidation coefficients and, on the other hand, it considers that primary and secondary settlements may occur in parallel.

### 4.1.1  Immediate settlement

Immediate settlements arise from instant loading and no variation of the clay volume. Thereafter, they are also known as undrained, elastic or distortional settlement (see scheme in Figure 4.2).

Usually, immediate settlements, $\Delta h_i$, are of small magnitude when compared to consolidation settlements, $\Delta h_a$, particularly in the case of large embankments (length and width), compared to the thickness of the soft clay layer.

### 4.1.2  Primary consolidation settlements

The magnitude of primary consolidation settlements must be calculated by dividing the foundation layer into sublayers according to available data from consolidation tests. The parameters can be obtained from the compression curve, as shown in Figure 4.3, which also shows the determination of the overconsolidation stress $(\sigma'_{vm})$ using the method proposed by Pacheco Silva (1970), which is more popular in Brazil than the Casagrande's method.

The equation for calculating primary consolidation settlements of a layer of clay of thickness $h_{clay}$, with effective vertical stress in situ $\sigma'_{vo}$ and overconsolidation stress $\sigma'_{vm}$ is:

$$\Delta h = h_{clay}\left[\frac{C_s}{1+e_{vo}}\cdot\log\left(\frac{\sigma'_{vm}}{\sigma'_{vo}}\right)+\frac{C_c}{1+e_{vo}}\log\left(\frac{(\sigma'_{vo}+\Delta\sigma_v)}{\sigma'_{vm}}\right)\right] \qquad (4.1)$$

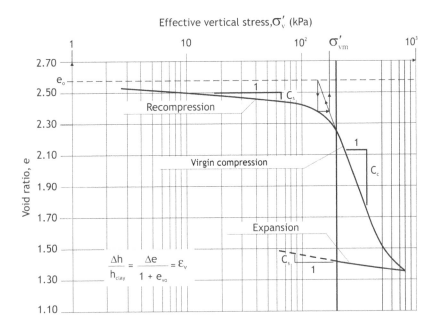

*Figure 4.3* Compressibility parameters from the compression curve – Method of Pacheco Silva (1970).

where $C_s$ and $C_c$ are the recompression and compression indexes, $e_{vo}$, is the *in situ* void ratio at the center of layer.

The increase in stress due to the loading of the embankment, $\Delta\sigma_v$ is calculated according to the geometry of the problem, as shown in Figure 4.4A:

$$\Delta\sigma_v = I \cdot (\gamma_{emb} \cdot h_{emb}) \qquad (4.2)$$

where $\gamma_{emb}$ is the unit weight of the embankment and $h_{emb}$, its height (thickness).

The influence factor I (Figure 4.4B) is provided by the Osterberg chart (Poulos, Davis, 1974). If the ratio b/z is high (greater than 3), i.e., wide embankments when compared with the thickness of the clay layer, the embankment is considered infinite and the factor I equals to 0.5 and $\Delta\sigma_v = 2 \times 0.5 \times \gamma_{emb} \times h_{emb}$, considering the symmetry of the embankment, which is the most common case. The $e_{vo}$ value obtained from the compression curve for $\sigma'_{vo}$ differs slightly from the $e_o$ value determined in laboratory (indicated in Figure 4.3), as this is higher due to the unloading of the sample during sampling.

As discussed in Chapter 3, poor quality samples cause alteration in the compression curves; it is recommended that the correction of the compression curve be performed according to the procedure proposed by Schmertmann (1955), shown in Figure 4.5. In the case shown in the figure, for a sample of poor quality there is a significant difference between the value of $e_{vo}$ and $e_o$, as discussed in section 2.7.2.

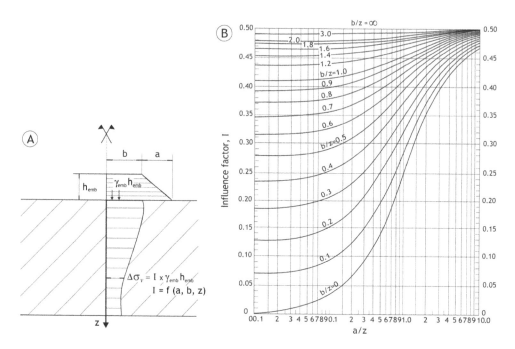

*Figure 4.4* (A) Parameters used for computing settlements; (B) Influence factor for trapezoidal loading (Poulos and Davis, 1974).

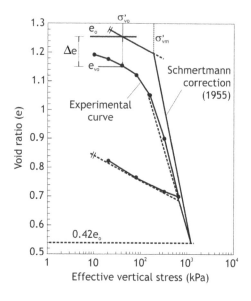

*Figure 4.5* Correction scheme for compression curve of the Schmertman consolidation test (1955).

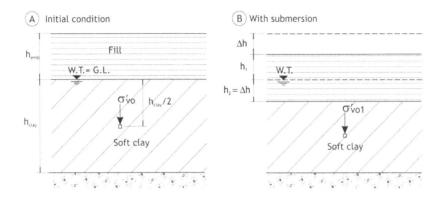

*Figure 4.6* Embankment submersion scheme: (A) beginning of loading; (B) after settlement Δh.

### Effect of embankment submersion

Calculation of settlements when considering the effect of submersion of an infinite embankment is iterative. Initially, the settlement value is calculated disregarding submersion of the embankment, which corresponds to the first iteration j, according to Equation (4.3), simplified for normally consolidated clay:

$$\Delta h_j = h_{clay} \frac{C_s}{(1 + e_{vo})} \cdot \log\left(\frac{(\sigma'_{vo} + \gamma_{emb} h_{emb})}{\sigma'_{vo}}\right) \tag{4.3}$$

Assuming the water table coincides with the ground level (Figure 4.6A), the height of the embankment $h_{emb}$ is then divided into $h_1$ and $h_2$ ($= \Delta h_j$), corresponding respectively to the non-submerged and submerged parts (submerged unit weight $= \gamma'_{emb}$), as shown in Figure 4.6B. The settlement calculated in the second iteration $\Delta h_{j+1}$ is given by:

$$\Delta h_{j+1} = h_{clay} \frac{C_s}{1 + e_{vo}} \cdot \log\left(\frac{(\sigma'_{vo} + \gamma_{emb} h_1 + \gamma'_{emb} h_2)}{\sigma'_{vo}}\right) \tag{4.4}$$

The calculations must be done until convergence is achieved, i.e., until the settlement $\Delta h_{j+1}$ of the current iteration $j + 1$ coincides with the settlement $\Delta h_j$ of the previous iteration j. This calculation procedure is valid for a layer of thickness equal to the clay. If there are two or more sublayers, the value $h_2$ (Figure 4.6B) should be equal to the sum of the settlements of all sublayers. The procedure described should be modified in case the water level does not coincide with the ground level.

Figure 4.7 shows an example of iterative calculation of a typically consolidated clay layer with $h_{clay}$ thickness equal to 5 m, 10 m and 15 m, and for embankment thickness equal to 3 m (Figure 4.7A). As expected, the difference between settlements with and without submersion increases with the thickness of the clay layer.

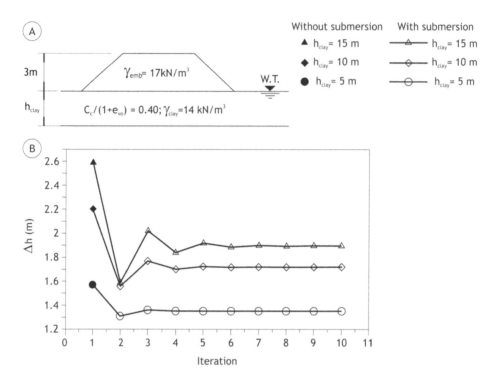

*Figure 4.7* Settlement considering submersion of embankment: (A) geotechnical model studied; (B) variation of settlement calculated according to iterations.

### Calculation for a fixed elevation of the embankment

The most common practical case is when settlement of an embankment must be stabilized at a fixed elevation, for example, embankments on bridge approaches and around piled buildings. The calculation process is iterative and in such cases, the settlement $\Delta h$ is used on both sides of the equation (Pinto, 2000) valid for a normally consolidated layer:

$$\Delta h = h_{clay}\frac{C_s}{1+e_{vo}} \cdot \log\left(\frac{(\sigma'_{vo} + \gamma_{emb}h_{emb} + \gamma'_{emb}\Delta h)}{\sigma'_{vo}}\right) \tag{4.5}$$

Figure 4.8 shows the variation of settlements for fixed embankment elevations for several clay layer thicknesses, assuming ground level at elevation $+0$ m. This figure shows that the necessary embankment thickness to reach a given elevation can be very high. For example, in the case of a 15 m thick clay layer and an embankment that must reach $+3$ m, the embankment thickness must be around 5 m, i.e., a 2 m settlement. For the same elevation, when the thickness of the clay increases, it is also necessary to increase the thickness of the embankment.

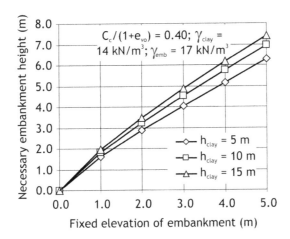

*Figure 4.8* Embankment height versus fixed elevation of embankment for different thicknesses of clay layers.

### Variation of primary consolidation settlements with time

The calculation of settlement variation over time can be done for two drainage conditions: one-dimensional drainage and radial drainage. The later is related with the use of prefabricated vertical drains (PVDs) installed to accelerate settlement, discussed in detail in Chapter 5.

### One-dimensional drainage – 1D

The calculation of settlements versus time in case of vertical drainage is based on Terzaghi's theory (Terzaghi, 1943). The calculation of settlement $\Delta h(t)$ at a given time t is performed by multiplying the primary consolidation settlement $\Delta h$ by the average degree of vertical consolidation $U_V$ (calculated according to Terzaghi theory), thus:

$$\Delta h(t) = U_v \cdot \Delta h \tag{4.6}$$

where $U_v$ is a function of the time factor $T_v$, according to Figure 4.9, for the conditions of drainage shown.

The time factor is a function of the coefficient of consolidation $(c_v)$ and the drainage distance $(h_d)$:

$$T_v = \frac{c_v t}{h_d^2} \tag{4.7}$$

The drainage distance is equal to the thickness $h_{clay}$ of the clay layer, in the case of single drainage, and equal to $h_{clay}/2$, when the layer has drainage on both sides. The choice of the design $c_v$ value is of great importance for an accurate estimation of the variation of settlements over time. In general, data from laboratory tests $(c_{vlab})$ and field tests $(c_{vpiez})$ are used for this purpose (see section 3.9). Back-analysis of settlement curves versus time supply $c_{vfield}$ data, which is very useful in the verification of the

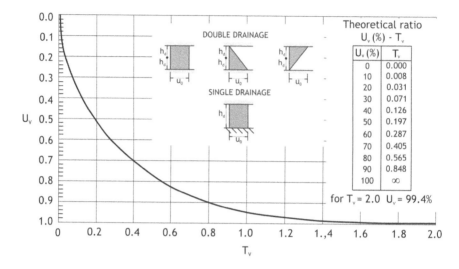

Figure 4.9 Variation of the degree of vertical consolidation with time factor.

design hypothesis. This is discussed in detail in Chapter 8. In usual loading cases, the following simplified equations can be used for calculation of $T_v$:

For $U_v \leq 60\%$

$$T_v = \left(\frac{\pi}{4}\right) U_v^2 \tag{4.8}$$

For $U_v \geq 60\%$

$$T_v = 1.781 - 0.933 \log(100 - U_v(\%)) \tag{4.9}$$

Figure 4.10 presents settlement versus time curves for a 3 m thick clay layer and embankment with elevation +3 m, with natural terrain elevation +0.5 m, that is, the minimum thickness necessary for an embankment without compensating the settlement is 2.5 m. In this case, the primary consolidation settlement for the fixed embankment elevation +3 m is equal to 0.95 m. Thus, it is necessary to use an embankment with thickness of at least 2.5 m + 0.95 m = 3.45 m to compensate for settlements. However, as long as stability is ensured, use of higher fills may be convenient to accelerate settlements. This procedure, detailed in Chapter 5, is known as temporary surcharge, as it is removed when the desired settlement and time are reached.

Calculations shown in Figure 4.10 were performed with a value of $c_v = 5.0 \times 10^{-8}$ m²/s for three different embankment thicknesses: 2.5 m, 4 m and 5 m, and the later thickness allows the fill to be removed in 22 months. In this case, the fill thickness to be removed is (5.0 − 0.95) − 2.5 = 1.55 m.

Figure 4.11 shows the period required for $U_v = 95\%$ versus the thickness of the clay layer. As the usual time required for stabilization of settlements is of a maximum of three years, PVDs are used for settlement acceleration in the case clay layers are thicker than 5 m with consolidation coefficients around $5.0 \times 10^{-8}$ m²/s.

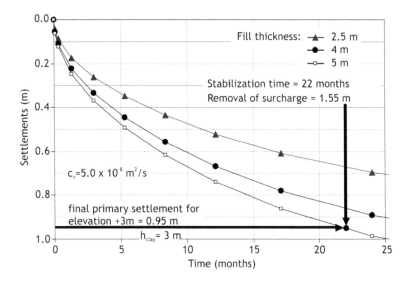

Figure 4.10 Variation of settlements over time for different embankment thicknesses.

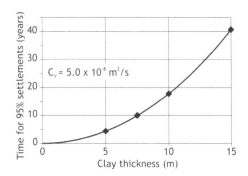

Figure 4.11 Stabilization time for 95% of settlements versus thickness of clay (double drainage).

### Non-instantaneous loading

Usually, the variation of settlement over time for instant loading of embankment (construction time zero) is computed and presented. However, in reality, the usual construction time $t_c$ for each loading phase is a few months. Thus, this changed settlement curve for $t_c$ different than zero might be presented according to the scheme shown in Figure 4.12, where two examples of settlement variation versus time for construction time of 30 and 360 days are shown. In this case, the method, as proposed by Terzaghi (1943) may be adopted which assumes that: a) settlement is equal to that which would occur if it were made instantaneously during half of the construction period $t_c$; b) settlements are proportional to loading. Thus, for time $t_c$, the settlement would be the same as in $t_c/2$. For times greater than $t_c$, the curve is obtained when corrected by the displacement of the instant settlement curve in a time $t_c/2$. For times

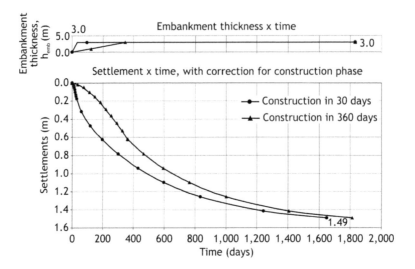

*Figure 4.12* Settlement versus time: influence of non-instantaneous loading.

$t < t_c$, this method assumes that the settlement at a time t for non-instantaneous loading is equal to the settlement that occurs at time t/2 and proportional to the applied load.

### 4.1.3  Secondary compression settlement

"Deformations at the end of primary consolidation and which cannot be attributed to small remaining excess pore pressure in the test specimen, are called secondary consolidations" (Martins, 2005). Considering that primary consolidation is related to dissipation of pore pressure, but the same does not happen with the "secondary compression," the latter name is adopted here instead of "secondary consolidation."

Researchers consider two hypotheses in compression:

- Hypothesis A, the traditional one, considers that secondary compression only happens after the end of the primary consolidation and does not depend on drainage conditions (Mesri, 1975; Jamiolkowski et al., 1985);
- Hypothesis B considers that the clay compression is due to structural soil viscosity, i.e., the vertical strain rate, and temperature. There are several approaches to this phenomenon (e.g., Taylor and Merchant, 1940; Mitchell, 1964; Kavazanjian and Mitchell, 1984; Leroueil et al., 1985; Martins and Lacerda, 1985; Leroueil and Watabe (2012).

*Traditional approach to secondary compression: Hypothesis A*

Secondary compression is seen in the laboratory, as shown in Figure 4.13A, where compression curves at the end of primary consolidation are shown as well as the traditional 24-hour curve (Martins, 2005). The traditional approach to secondary compression assumes that it occurs after the end of the primary consolidation, and for

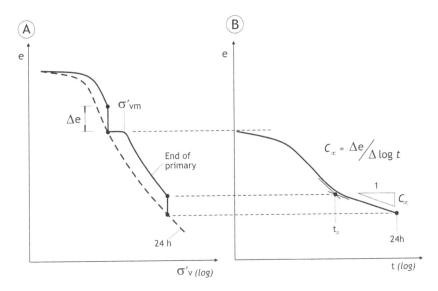

*Figure 4.13* Traditional approach of secondary compression: (A) compression curves at the end of primary and after 24 hours; (B) variation of void ratio of a loading stage (Martins, 2005).

each increase of applied vertical stress, the coefficient of secondary compression $C_\alpha$ is calculated, as shown in Figure 4.13B.

As shown in Figure 4.13, the variation of the primary consolidation settlements over time is calculated according to Equation (4.1) until the end of the primary consolidation (time $t_p$ for instance corresponding to $U = 95\%$). Then, the secondary compression settlements $\Delta h_{sec}$ start, which may be calculated according to:

$$\Delta h_{sec} = \frac{C_\alpha \cdot h_{clay} \cdot \log \frac{t}{t_p^*}}{(1 + e_{vo})} \tag{4.10}$$

The total settlement over time, according to this approach, is calculated as:

$$t \leq t_p \Rightarrow \Delta h(t) = \overline{U}(T_v) \cdot \Delta h_a \tag{4.11}$$

$$t = t_p^* \Rightarrow \Delta h(t_p^*) = \Delta h_a \tag{4.12}$$

$$t \geq t_p^* \Rightarrow \Delta h(t) = \Delta h_a + \frac{C_\alpha \cdot h_{clay}}{1 + e_{vo}} \log\left(\frac{t}{t_p^*}\right) \tag{4.13}$$

where $t_p^*$ is shown in Figure 4.14, for field analysis.

This approach is easy to use but the magnitude of secondary settlements obtained over time using this method is questionable, since it considers the fact that the secondary compression is endless as $C_\alpha$ is considered constant, i.e., the voids ratio would go towards negative values with time, which is not possible physically.

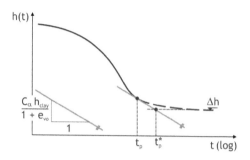

*Figure 4.14* Settlement curve versus time (Martins, 2005).

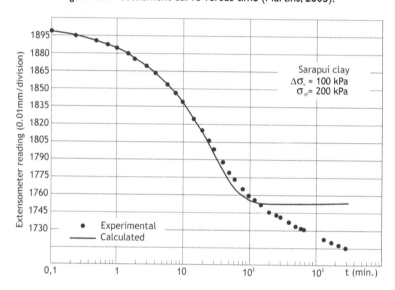

*Figure 4.15* Comparison between theoretical and experimental curves (Feijó, 1991).

## Influence of the relation $\Delta\sigma_v/\sigma_v$ on secondary compression

Leonards and Girault (1961) noted that the larger the value of the ratio $\Delta\sigma_v/\sigma_v$, the more the consolidation curve approaches Terzaghi's theoretical curve. Figure 4.15 shows experimental results (Feijó, 1991) that confirm this statement and this is the reason for which the ratio $\Delta\sigma_v/\sigma_v = 1$ is used in the laboratory. The smaller the loading ratio, the larger the contribution of the secondary compression settlement and the more the experimental curve will differ from the theoretical curve.

## Theory of Taylor and Merchant (1940) – Hypothesis B

Taylor and Merchant's theory (1940) is based on the structural viscosity of the soil. It considers the influence of the ratio $\Delta\sigma_v/\sigma_v$ and it predicts that secondary compression in the field occurs in parallel with primary consolidation (hypothesis B). Thus, it is a step beyond the Terzaghi and Fröhlich (1936) theory. Figure 4.16 shows the solution of the equations for this theory (Martins, 2005) in terms of the average degree of

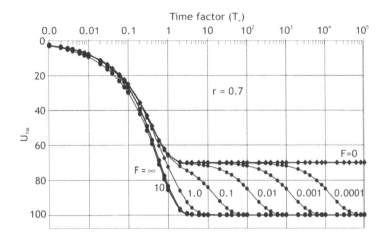

*Figure 4.16* Average percentage of consolidation $U_{TM} \times T_v$ – Theory of Taylor and Merchant (Martins, 2005).

consolidation $U_{TM}$ versus the time factor $T_v$ for the parameter $r = 0.7$, which is defined from the relation between the primary and total settlements $r = \Delta h_a/(\Delta h_a + \Delta h_{sec})$. F is the additional calculation parameter necessary when using this theory and is defined by:

$$F = \frac{\mu h_d^2}{r c_v} \tag{4.14}$$

where $\mu$ is the value of soil viscosity.

The value of F for field conditions is usually around 10 and, as shown in Figure 4.16, the solution for this value of F is close enough to the solution for $F = \infty$ (Martins, 2005). Thereby, the trace line of the field settlement curve according to the theory consists in estimating the value of r and the line of the $U_{TM}$ curve versus T for $F = \infty$, thus allowing calculation of the settlement at any time t, by multiplying the total settlement by the $U_{TM}$ value.

Calculation of the magnitude of total settlements (primary + secondary) using the curve $U_{TM} \times T_v$ requires estimating the $\Delta h_{sec}$ value, which is shown below, based on studies with Brazilian clays.

Remy et al. (2010) applied the theory of Taylor and Merchant (1940) in back-analysis of settlements of two test embankments with vertical drains. There was good correlation between the values of consolidation coefficient measured in the laboratory and those obtained from their back-analysis, for which they adopted $c_v = c_h$.

### Estimate of the secondary compression settlement $\Delta h_{sec}$ according to Martins (2005)

Based on laboratory data, Martins (2005) proposed that the maximum secondary consolidation settlement corresponds to the variation of vertical strain from the 24h

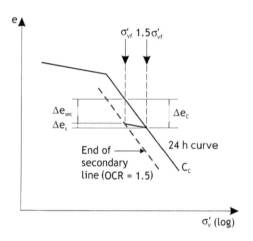

*Figure 4.17* Construction of end of secondary line.

curve to a line of OCR $= 1.5$, for a given effective vertical stress $(\sigma'_{vf})$ acting on a soft clay, as shown in Figure 4.17.

The end line of secondary compression in diagram $e \times \sigma'_v$ may be obtained in the laboratory, generating an OCR $= 2$, from the end of primary, or an OCR $= 1.5$ from the 24-hour line. This is the reference line for the calculation of consolidation settlements. Thus, the calculation of the secondary compression settlement may be associated with the calculation of primary consolidation settlements, assuming compression until stress $1.5\sigma'_{vf}$, followed by unloading until $\sigma'_{vf}$.

Thus, for $CR = \dfrac{C_c}{1 + e_{v0}}$, and assuming that $\dfrac{C_S}{C_c} = 0.15$:

$$\Delta h_{sec} = h_{clay} CR \log\left(\frac{1.5\sigma'_{vf}}{\sigma'_{vf}}\right) - h_{clay}(0.15 CR) \cdot \log\left(\frac{1.5\sigma'_{vf}}{\sigma'_{vf}}\right) \tag{4.15}$$

$$\frac{\Delta h_{sec}}{h_{clay}} = 0.15 CR \tag{4.16}$$

Thereby assuming a CR value of 0.40, common in very soft clays, one obtains $\Delta h_{sec} = 0.06\, h_{clay}$. Thus, for clays of thickness $h_{clay} = 10\,m$, the estimated secondary settlement is $\Delta h_{sec} = 0.60\,m$. Using $h_{clay}$ fixed height $= +3\,m$ Figure 4.8 shows a primary consolidation settlement $\Delta h_a = 1.5\,m$ (thickness of embankment – fixed height $= 4.5\,m - 3.0\,m$, see Figure 4.8). These values result in $r = 1.5/(0.6 + 1.5) = 0.7$, confirming the importance of the secondary compression settlements compared to the primary consolidation settlements for clays of high CR value.

## 4.2  STAGED EMBANKMENT SETTLEMENTS

If the embankment is not stable to build in a single stage, alternative building methods must be used, one of which is to build the embankment in stages (Almeida, 1984; Ladd,

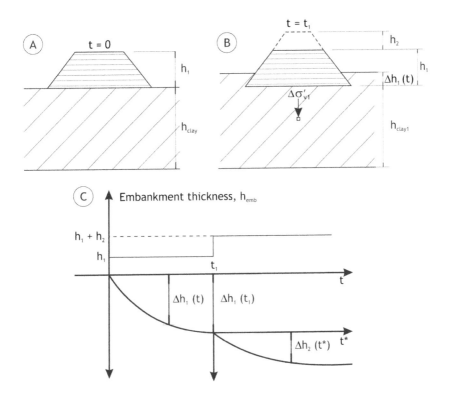

*Figure 4.18* Scheme of an embankment built in two stages.

1991), allowing time for the foundation's soft soil to gain strength before adding the next layer.

Two types of calculations are relevant when building the embankment in stages: calculation of the settlement variation over time described below, and stability calculations, described in Chapter 6.

Stage construction is shown in Figure 4.18 for the case of two stages: the first, for $t = 0$ (Figure 4.18A), and the second, for $t = t_1$ (Figure 4.18B). The most common is the use of two or three construction stages.

Calculation of settlements of staged constructed embankment follows the usual procedure; however, stabilization of 95% of the settlements of one stage before adding the next stage is not usually adopted, because this would require a long time. Calculation of settlements for more than one stage must be performed according to the following steps:

1 Calculating the total $\Delta h_1$ settlement correspondent to height of embankment $h_1$
   This calculation is done in the conventional way using the equations shown throughout this chapter. Assuming, for simplification, that the clay is normally

consolidated ($\sigma'_{vm} = \sigma'_{vo}$), and neglecting embankment submersion effects, the final settlement of the first stage of loading is:

$$\Delta h_1 = h_{clay1} \cdot [C_c/(1 + e_{vo})] \cdot \log[(\sigma'_{vo} + \gamma_{emb} \cdot h_1)/\sigma'_{vo}] \qquad (4.17)$$

2   Calculating the settlement variation $\Delta h_1(t) = \Delta h_1 \cdot U_1$ for each time t, through $t_1$, corresponding to the beginning of the second stage. Usually $U_1(t_1) \geq 60\%$, is adopted and the value of $U_1$ used is the one corresponding to the vertical drainage without drains, or radial or combined drainage when using PVDs.
   In case of thicker layers, staged construction is usually associated with the use of PVDs (radial drainage), which allow the clay to gain strength faster. However, loading in stages can also be associated to solely vertical drainage, in case of thinner layers.

3   For settlement calculation after time $t_1$, values of each sublayer must be used according to Figure 4.18B, namely:
   a.   Thickness of layer:

$$h_{clay1} = h_{clay} - \Delta h_1 \cdot U_1 \qquad (4.18)$$

   where $U_1 = U_1(t_1)$.
   b.   Effective vertical stress in time $t_1$:

$$\sigma'_{v1} = \sigma'_{vo} + U_1 \cdot (\gamma_{emb} \cdot h_1) \qquad (4.19)$$

4   Settlement after the time $t_1$ comes from 2 portions:
   a.   Settlement that has yet to occur, referring to the height of embankment $h_1$, corresponding to the increment of effective stress, referring to pore pressures yet to be dissipated:

$$\Delta\sigma'_{v1} = (1 - U_1) \cdot (\gamma_{emb} \cdot h_1) \qquad (4.20)$$

   b.   Settlement referring to the height of embankment $h_2$, corresponding to the increment of effective stress in stage 2:

$$\Delta\sigma'_{v2} = \gamma_{emb} \cdot h_2 \qquad (4.21)$$

Assuming the clay is normally consolidated ($\sigma'_{v1} > \sigma'_{vm}$), total settlement for second stage will be:

$$\Delta h_2 = h_{clay1} \cdot [C_c/(1 + e_{vo})] \cdot \log[(\sigma'_{v1} + \Delta\sigma'_{v1} + \Delta\sigma'_{v2})/\sigma'_{v1}] \qquad (4.22)$$

5   Calculation of settlement variation $\Delta h_2(t^*) = \Delta h_2 \cdot U$ for each time $t^*$, considering $t = t_A$ the new origin of time $t^* = 0$, as shown in Figure 4.18C.

A value for $c_v$ may be adopted for stage 2 that is different from the corresponding value at stage 1 (e.g., Coutinho, Almeida and Borges, 1994), as the coefficient of consolidation decreases with increase of effective stress and also as the clay goes from overconsolidated to normally consolidated. However, as the overconsolidated condition is usually reached at the end of construction (Leroueil et al., 1978; Leroueil and

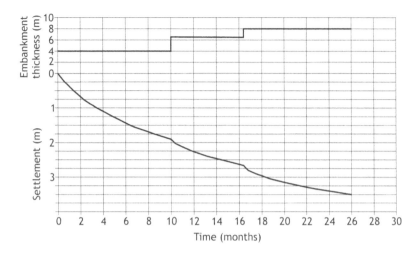

*Figure 4.19* Evolution of settlements over time.

Tavenas, 1986), in general the normally consolidated value of $c_v$ is utilized, which is a bit more conservative in terms of construction periods. In general, the ground level is superficial, thus it is necessary to consider the submersion in a the calculation of $\Delta h_1$ and $\Delta h_2$.

In case of a third stage, items 3 to 5 should be repeated.

Settlement calculation for staged loading may be done quickly by using spreadsheets. Figure 4.19 shows an example of a predicted settlement versus time. Curve for an embankment built in three stages in extremely soft clay.

## 4.3  PREDICTION OF HORIZONTAL DISPLACEMENTS

The oedometer consolidation test simulates the behavior of a clay soil element that, when loaded, has zero horizontal displacements, such as the soil element at the center of an embankment. However, at the edges of the embankment, where there is no lateral confinement, horizontal displacements ($\delta_h$) may be of practical importance. This is the case for structures adjacent to the embankment, where it is necessary to estimate the horizontal displacements as well.

The displacement magnitude under an embankment is a result of the stress paths. Considering an element of clay soil located under the center line of an embankment, with initial stress $I_1$ (Figure 4.20A), with the construction of an embankment in one stage, the stress path is close to $K_0$ line ($I_1 - C_1 - E_1$), in the overconsolidated domain, with relatively small displacements. In this domain, the magnitude of settlements is high, but such settlements happen slowly, as the $c_v$ values are smaller. However, in the case of proximity to failure, the horizontal displacements increase rapidly (see Chapter 8).

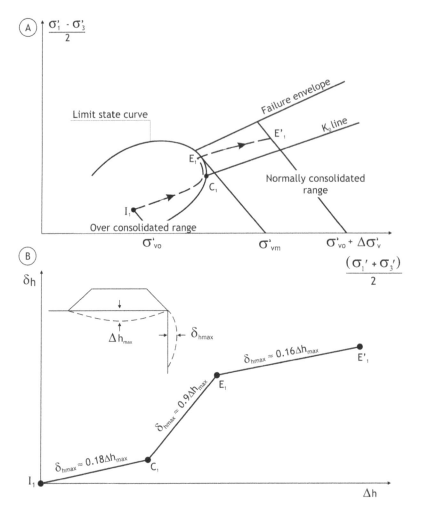

Figure 4.20 Estimate of the relation between maximum settlement under the center of an embankment and maximum horizontal displacement of the edge of the embankment (Tavenas, Mieussens and Bourges, 1979).

Maximum horizontal displacements ($\delta_{hmax}$) may be estimated from the empirical correlation with the maximum settlements ($\Delta h_{max}$) measured in the centerline of the embankment, through the method proposed by Tavenas, Mieussens and Bourges (1979), which correlated $\delta_{hmax}$ and $\Delta h_{max}$ (Figure 4.20B) through:

$$DR = \frac{\Delta h_{max}}{\delta_{h\,max}} \qquad (4.23)$$

For embankments built in one stage, these authors concluded, from the analyses of 15 embankments with slopes of about 1.5 a 2.5(H):1.0(V), built in deposits

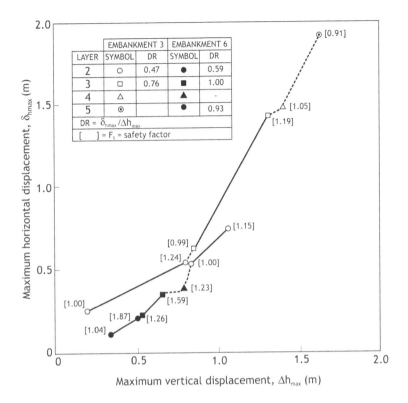

*Figure 4.21* Maximum settlements versus maximum horizontal displacements for Embankments 3 and 6 (Almeida, 1984).

with OCR $< 2.5$ and without vertical drains, that there are two successive behavior conditions during the loading phase:

1   Partially drained: during the initial loading phase – because of the initial $c_v$ of the overconsolidated soil – horizontal displacements happen quickly and are initially much smaller than vertical displacements, which results in the correlation:

$$DR = \frac{\Delta h_{max}}{\delta_{h\,max}} = 0.18 \tag{4.24}$$

2   Undrained: as effective stresses increase with loading, the clay layer moves into the normally consolidated range, and the horizontal displacements become of the same magnitude as the vertical displacements, resulting in the correlation:

$$DR = \frac{\Delta h_{max}}{\delta_{h\,max}} = 0.9 \tag{4.25}$$

3   In the consolidation stage following construction, the authors concluded that the horizontal displacement continues to increase linearly with settlement, based on the analyses of twelve embankments, resulting in the correlation:

$$DR = \frac{\Delta h_{max}}{\delta_{h\,max}} = 0.16 \qquad (4.26)$$

For more complex cases, according to Ladd (1991), the correlations proposed by Tavenas, Mieussens and Bourges (1979) have limited applicability to the conditions of the analyzed cases. Ladd (1991) emphasizes that the significant deviations of the patterns described here may be found in case of PVDs and more so in the case of loading in stages and foundations with large areas on yield conditions.

Results achieved by Almeida (1984) confirmed Ladd's findings as shown in Figure 4.21, which shows diagrams of maximum vertical displacement versus horizontal displacement for Embankments 3 and 6 (see item 7.2.6), both built in stages, the first on virgin foundation and the latter on a foundation supported by granular columns. It is worth noting that the DR values resulting from the consolidation phases in each stage of both embankments are far superior to the one obtained from Equation (4.26).

## 4.4   FINAL REMARKS

The expected settlements for embankments on very soft compressible deposit are generally quite high. Brazilian soft clays are in general lightly overconsolidated and present compression ratio values of CR above 0.25 (see Annex). In clays with very high compressibility as for instance those found in Barra da Tijuca (RJ), vertical strains caused by the construction of the embankment to achieve a fixed height in the order of 3 m may be about 30% (Almeida et al., 2008c).

The analyses presented in this chapter are current analyses to be performed in engineering practice. However, more sophisticated analyses regarding the behavior of soft soils using analytical or numerical methods are becoming quite common more recently (e.g., Chai and Carter, 2011).

The magnitude of settlements and their progress over time, as well as the post-construction settlements, must be considered when choosing the construction method to be adopted, which also depends on the objectives of the area to be used. Generally, in cases of residential or commercial buildings, or railroads embankments, post- construction settlements are not acceptable. However, in some cases of industrial sites and road embankments some amount of post-construction (either primary or secondary) settlements may be tolerated.

Because of the possible discrepancies between expected behavior and actual behavior, it is essential that embankments on soft clays be monitored (see Chapter 8), so that adjustments can be made during the construction period.

# Chapter 5

# Acceleration of settlements: use of vertical drains and surcharge

Vertical drains promotes faster settlements due to the decrease of the drainage path within the compressible soil mass to about half the horizontal distance between drains.

Temporary surcharge also accelerates settlements related to the primary consolidation and it decreases post-construction settlements. Thus, the combination of prefabricated vertical drains and temporary surcharging fully explores the benefit of rapid consolidation. Vertical drains and surcharge are widely used in the construction of embankment for roads, railroads, airports, ports and storage areas in general.

## 5.1 EMBANKMENTS ON VERTICAL DRAINS

Vertical sand drains were first used in the late 1920s, in California, United States, and the use of prefabricated vertical drains (PVD) started in the 1970s. A PVD consists of a PVC core with a geotextile filter around it.

PVDs should have high mechanical resistance, which guarantees their structural integrity during installation. This allows them to resist both driving stresses and forces from horizontal and vertical deformations during the consolidation of the soil mass. In contrast, traditional sand drains are very susceptible to damage during their implementation and operation. In very soft clays, sand drains may suffer shear failure that causes them to become inoperative.

The direction of the water flow within the soil mass goes from predominantly vertical to mostly horizontal (radial) with the installation of vertical drains. The water collected by vertical elements is directed to the natural ground surface, to the drainage blanket, which must have sufficient thickness and inclination to be released to the atmosphere by means of gravity or by pumping, depending on the length of the blanket. Horizontal drains can be installed inside the blanket (Figure 5.1A) to facilitate water release (see section 5.3). At the end of the installation, vertical drains may be covered by the drainage blanket or by an embankment, as seen in the outline of Figure 5.1B.

Figure 5.2 illustrates the advantage of using vertical drains to accelerate the settlements of an embankment on soft soil, when comparing the evolution of settlements with time of an embankment without drains on a thick layer of soft soil.

The theoretical and practical aspects related to the use of vertical drains are addressed by Magnan (1983) and Holtz et al. (1991) and are summarized below.

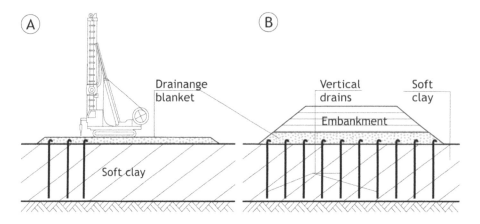

*Figure 5.1* Outline of PVD installation in a soft clay layer underlying an embankment.

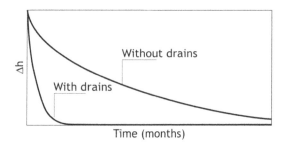

*Figure 5.2* Settlements versus time: with and without drains.

## 5.2 VERTICAL DRAINS

### 5.2.1 Theoretical solutions

The consolidation of a compressible soil layer, considering only a vertical water flow, one-dimensional (1D), is given by the differential equation:

$$\frac{\partial u}{\partial t} = c_v \frac{\partial^2 u}{\partial z^2} \tag{5.1}$$

Three-dimensional consolidation (3D), considering flow in directions x, y and z, is governed by the equation:

$$\frac{\partial u}{\partial t} = c_h \left[ \frac{\partial^2 u}{\partial x^2} + \frac{\partial^2 u}{\partial y^2} \right] + c_v \frac{\partial^2 u}{\partial z^2} \tag{5.2}$$

Considering isotropic conditions in directions x and y, the coefficient of horizontal consolidation is given by:

$$c_h = \frac{k_h(1 + e_{vo})}{a_v \gamma_w} \qquad (5.3)$$

where:
$e_{vo}$ – initial void ratio corresponding to the effective vertical stress in situ;
$a_v$ – coefficient of compressibility;
$x, y, z$ – coordinates of a soil mass point;
$u$ – excess pore pressure;
$c_v$ and $c_h$ – consolidation coefficients for vertical and horizontal drainage, respectively;
$k_v$ and $k_h$ – vertical and horizontal coefficients of permeability, respectively;
$\gamma_w$ – specific weight of water.

Equation (5.2) represents consolidation from a combined vertical and horizontal flow occurring, for instance, on the edges of an embankment without drains. When using vertical cylindrical draining elements, Equation (5.2) may be transformed to cylindrical coordinates:

$$\frac{\partial u}{\partial t} = c_h \left[ \frac{1}{r} \frac{\partial u}{\partial r} + \frac{\partial^2 u}{\partial r^2} \right] + c_v \frac{\partial^2 u}{\partial r^2} \qquad (5.4)$$

where r is the radial distance measured from the drainage center to the point considered.

## 5.2.2 Consolidation with purely radial drainage

If vertical drainage in the soil mass is disregarded, then the pure radial drainage is given by the Equation:

$$\frac{\partial u}{\partial t} = c_h \left[ \frac{1}{r} \frac{\partial u}{\partial r} + \frac{\partial^2 u}{\partial r^2} \right] \qquad (5.5)$$

Barron (1948) solved Eq. (5.5) for a cylinder of soil with cylindrical vertical drain, for the condition of vertical equal strain, thus obtaining the average rate of consolidation of the layer, $U_h$:

$$U_h = 1 - e^{-[8T_h/F(n)]} \qquad (5.6)$$

where:

$$T_h = \frac{c_h \cdot t}{d_e^2} \qquad (5.7)$$

$T_h$ – time factor for horizontal drainage;

$$F(n) = \frac{n^2}{n^2 - 1} \ln(n) - \frac{3n^2 - 1}{4n^2} \cong \ln(n) - 0.75 \qquad (5.8)$$

$$n = \frac{d_e}{d_w} \qquad (5.9)$$

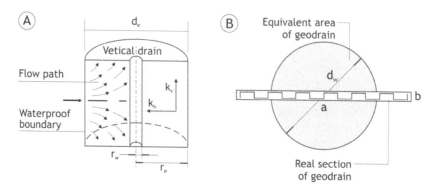

*Figure 5.3*   Geometric parameters of drains: (A) area of drain influence and detail of unit cell; (B) detail of equivalent section of a PVD.

where:

$d_e$ – diameter of influence of a single drain (Figure 5.3A);

$d_w$ – diameter of drain or equivalent diameter of a geodrain with rectangular section (Figure 5.3B);

Barron (1948) also solved the equation for free strain condition. In this case, vertical free strains are allowed on the surface of the influence cylinder of one drain as consolidation occurs. This solution is presented in terms of Bessel functions and for values of n > 5 (the case of PVDs) the two solutions are very similar. For this reason, the solution for the equal strain condition is generally used because of its simplicity.

The value of $c_h$ that should be used can be defined by laboratory or in situ tests, as described in detail in section 3.4.6.

It is worth mentioning that the values to be used for the coefficients of consolidation are the ones related to the stress range relevant to the problem. The use of $c_v$ or $c_h$ for normally consolidated conditions is appropriate in most cases of very soft soils.

### 5.2.3   Diameter of influence and equivalent diameter of PVDs

Figure 5.3 shows the geometric parameters of the drains to be discusses here. The diameter of influence of a drain $d_e$ (see Figure 5.3A) is a function of drain spacing and its configuration in a triangular or square pattern with spacing equal to l. For the square pattern, shown in Figure 5.4A, for equal areas of square and circle one obtains:

$$l^2 = \frac{\pi d_e^2}{4} \quad \text{and} \quad d_e = l\sqrt{\frac{4}{\pi}} \qquad (5.10)$$

Thus obtaining the diameter of influence of a square mesh:

$$d_e = 1.131 \qquad (5.11)$$

And for the triangular pattern represented in Figure 5.4B, when the area of the equivalent circle is equated to the hexagon:

$$\frac{\pi \cdot d_e^2}{4} = \frac{\sqrt{3}}{2}l^2 \quad \text{and} \quad d_e = \sqrt{\frac{2}{\pi}\sqrt{3}\,l} \qquad (5.12)$$

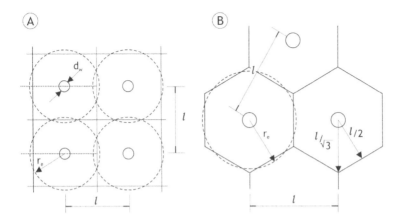

*Figure 5.4* Geometric data of vertical drains: (A) square pattern; (B) triangular pattern.

That is, the diameter of influence for triangular pattern is given by:

$$d_e = 1.05l \tag{5.13}$$

PVDs are generally rectangular and their dimensions a and b (Figure 5.3B) are typically 10 cm and 0.5 cm, respectively. The PVD is modeled by an equivalent diameter ($d_w$) which, according to Hansbo's proposal (1979), must have the same perimeter of a circular drain. Thus, the equivalent diameter of a PVD is represented by:

$$d_w = \frac{2(a + b)}{\pi} \tag{5.14}$$

Subsequent studies (Atkinson and Eldred, 1981; Rixner, Kreaemer and Smith, 1986) recommend the equivalent diameter of the PVD to be:

$$d_w = \frac{(a + b)}{2} \tag{5.15}$$

In practice, Eq. (5.14) is used more often than Eq. (5.15). The difference in using one or the other to calculate the spacing of the drains is negligible because of the wide variation of $c_h$ throughout the layer profile. Hansbo (2004) presented equivalent diameter values (according to Eq. 5.14) varying between 62 mm to 69 mm for fifteen PVDs available in the market and an average value of 65 mm.

### 5.2.4  Consolidation with combined radial and vertical drainage

When using a vertical drain in layers that have relatively small thickness vertical drainage must be considered in addition to radial drainage. This combined drainage was addressed theoretically by Carrillo (1942) who solved Eq. (5.4) to obtain the average percentage of combined consolidation U:

$$(1 - U) = (1 - U_v) \cdot (1 - U_h) \tag{5.16}$$

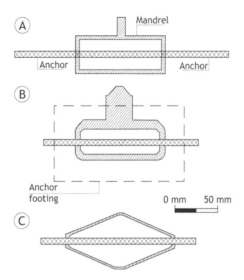

*Figure 5.5* Cross sections of mandrel and anchor installation (Saye, 2001).

### 5.2.5   Influence of smear in PVD performance

The installation process consists of positioning the drain in the interior of a hollow vertical metal rod known as mandrel. The PVD is then connected to an anchor, which is lost during installation (see detail in Figure 5.5).

The disturbance of the clay (smear effect) around the PVD caused by the installation process decreases the permeability of the soil around it and therefore reduces the consolidation rate and the efficiency of the PVD and furthermore, increases the magnitude of total settlements (Saye, 2001). Therefore, installation of PVDs should be hydraulic rather than by impact or vibration, which disturbs a larger volume of soil.

### 5.2.6   Influence of mandrel size on soil disturbance

The mandrel must have the smallest possible cross section to minimize disturbance. The outer area of the mandrel should be around $70\,cm^2$ ($6\,cm \times 12\,cm$) to fulfill structural requirements for the installation equipment on very soft soils. If the very soft clay layer contains compacted sand layers or shells, or if its thickness is greater than $15\,m$, it may be necessary to use a mandrel with external support, which may lead to greater disturbance (Sandroni, 2006). Figure 5.5 shows examples of mandrel and anchor footing.

Saye (2001) showed that the size of the mandrel and the anchor are responsible for disturbances and defined the modified spacing ratio $n' = d_e/d_m^*$, where $d_m^*$ is the equivalent diameter of the mandrel defined by its perimeter divided by $\pi$. The author suggested a minimum distance between PVDs around $n' = 7$, for a consolidation coefficient $c_h/c_v = 1.0$, valid for isotropic soft clays, and $n' = 10$ for clays with consolidation coefficient ratio $c_h/c_v$ of about 4.0. In other words, for PVDs that are very close to each other, the permeability reduction in the disturbed area can be excessive and reducing

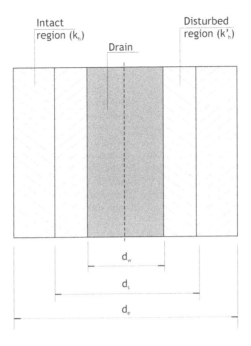

*Figure 5.6* Disturbed region around a vertical drain.

the spacing becomes a disadvantage. This minimum distance is due to the sensitivity of the soil and the geometry of the anchor footing and mandrel. Saye (2001) examined the cases of constructions with different spacing and recommended a minimum distance l between drains equal to 1.75 m for a case in which anchor footing with an area of 181 cm$^2$ was used.

More recent studies by Smith and Rollins (2009), for an installation footing area of 116 cm$^2$, showed a minimum value between 0.9 m and 1.22 m. Smith and Rollins (2009) recommended a minimum distance between PVDs of $n' = 8$ for $c_h/c_v$ of about 4.0.

### 5.2.7  Parameters for consideration of disturbance (smear)

Figure 5.6 shows an outline of the disturbed area around a PVD. When considering the disturbance, one should add to the value F(n), in Equation (5.8) the value $F_s$ corresponding to smear (Hansbo, 1981):

$$F_S = \left(\frac{k_h}{k'_h}\right) \ln\left(\frac{d_s}{d_w}\right) \tag{5.17}$$

where $k_h$ is the coefficient of permeability of the intact area and $k'_h$ is the coefficient of permeability of the disturbed zone, which diameter $d_s$ may be assumed equal to two

*Table 5.1* Dimensions and permeability ratio for the disturbed zone (adapted from Indraratna et al., 2005).

| Reference | $d_s/d_m$ | $k_h/k_h'$ | Observations |
|---|---|---|---|
| Barron (1948) | 1.6 | 3 | Assumed |
| Hansbo (1979) | 1.5~3 | – | Based on available literature at the time |
| Hansbo (1981) | 1.5 | 3 | Assumed in case study |
| Hansbo (1997); Basu, Prezzi, Madhav (2010) | 2~3 | 3~4 | Recommendations for design |
| Bergado et al. (1991) | 2 | 1* | Laboratory tests and back analysis of embankments on soft clay in Bangkok |
| Onoue et al. (1991) | 1.6 | 3 | Test interpretations |
| Almeida et al. (1993) | 1.5~2 | 3~6 | Based on authors' experience |
| Indraratna & Redana (1998) | 4~5 | 1.15* | Laboratory tests for Sydney clay |
| Hird et al. (2000) | 1.6 | 3 | Recommendations for design |
| Xiao (2000) | 4 | 1.3 | Laboratory tests for kaolin clays |

*$k_h/k_v$

times the equivalent diameter of the mandrel $d_m$ (Hansbo, 1987), i.e., $d_s = 2d_m$, where $d_m$ is given by:

$$d_m = \sqrt{\frac{4}{\pi} w \cdot l}$$

(5.18)

where w and l are the dimensions of a rectangular mandrel (Bergado et al., 1994).

The parameters for the effect of the disturbance resulting from installation ($k_h'$ and $d_s$) may have great influence on pore pressure dissipation if the installation process causes excessive disturbance. In the absence of data regarding $k_h'$ Hansbo (1981) recommends adopting:

$$\frac{k_h}{k_h'} = \frac{k_h}{k_v}$$

(5.19)

where $k_h$ is the coefficient of permeability of the area that remains intact and $k_h'$ is the coefficient of coefficient of permeability of the disturbed zone.

The coefficient of permeability relation $k_h/k_v$ generally varies between 1.5 to 2 for Brazilian soft clays (Coutinho, 1976), but can reach values of about 15 for stratified clays (Rixner, Kreaemer and Smith, 1986).

Recommendations on the effect of disturbance are summarized in Table 5.1. The range of values presented in the literature for the geometry of the disturbed area indicates $d_s/d_m$ between 1.5 and 5, with an average value of $d_s/d_m = 2.3$. Jamiolkowski and Lancellotta (1981) suggested, for design purposes, $d_s/d_m$ in the range 2.5–3.0. The disturbed area usually has lower coefficient of permeability than the region which remains intact. The studies suggest range of values for $k_h/k_h'$ between 1 and 6, with an average value $k_h/k_h' = 2.5$.

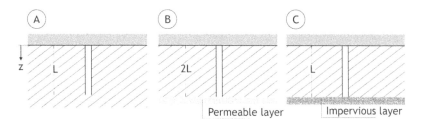

*Figure 5.7* Characteristic length of a drain.

## 5.2.8  The effect of well resistance

The discharge capacity of a drain is related to the drain cross section. This area decreases with the increase of the soil horizontal stresses, bending of PVDs due to settlement of the soft clay layer as well as clogging of the drains. In other words, PVDs do not have infinite coefficient of permeability, as admitted by Barron (1948) in deriving Eq. (5.6). Orleach (1983) proposed that the hydraulic resistance of PVDs should be given by the following equation derived from Hansbo (1981):

$$W_q = 2\pi \frac{k_h}{q_w} L^2 \tag{5.20}$$

where $q_w$ is the discharge capacity or flow of the PVD measured by testing for a unit gradient $i = 1.0$; and L is the characteristic length of the PVD, defined as the length of the PVD when drainage occurs only at one end (Figure 5.7A,C), and as half of it when drainage occurs at both ends (Figure 5.7B).

If $W_q < 0.1$, the hydraulic resistance of the PVD can be disregarded, otherwise Hansbo (1981) recommends adding to the value of F(n), Eq. (5.8), the value of $F_q$, defined by

$$F_q = \pi z(L - z)\frac{k_h}{q_w} \tag{5.21a}$$

Since $F_q$ is a function of z, then $U_h = f(z)$. Therefore, an average value of $U_h$ is adopted. For the specific case of double draining layer (Figure 5.7B), the equation to be used is:

$$F_q = \frac{2k_h \cdot \pi l^2}{3q_w} \tag{5.21b}$$

Long drains (greater than 20 m) with small discharge capacity may influence the well resistance (Jamiolkowski et al. 1983).

Most PVDs available on the market have enough discharge capacity ($q_w >$ 150 m$^3$/year), thus this issue may be negligible in a project (Hansbo, 2004). The recently launched PVDs with integrated filter and core, also known as integrated PVDs (Liu and Chu, 2009), offer more discharge capacity than conventional drains.

## 5.2.9 Specification of PVD

The main characteristic a PVD should have is to be more permeable than the soil and to maintain this permeability throughout its service life. To do so, a PVD is basically determined by $q_w$ and by the coefficient of permeability of the filter. Mechanic resistance and flexibility are also important characteristics, because the PVD must withstand driving operations and deformations imposed by the soil during consolidation.

Bergado et al. (1994) and Holtz, Shang and Bergado (2001) proposed that $q_w$ should not be less than a value between ($\sim$100 and 150 m³/year), when measured under a unitary hydraulic gradient and under maximum effective lateral confining stress acting in the field. The coefficient of permeability of the filter should be generally greater than ten times the soil's coefficient of permeability, by adopting the largest possible filtering opening of the geotextile, based on the soil retention criteria described by:

$$\frac{O_{90}}{D_{50}} < 1.7 \text{ to } 3 \quad \text{(Schober and Teindel, 1979)} \tag{5.22}$$

$$\frac{O_{90}}{D_{85}} < 1.3 \text{ to } 1.8 \quad \text{(Chen and Chen, 1986)} \tag{5.23}$$

$$\frac{O_{50}}{O_{50}} < 10 \text{ to } 20 \quad \text{(Chen and Chen, 1986)} \tag{5.24}$$

where:
$O_{90}$ – filter opening of geotextile, defined as the diameter of the biggest soil grain capable of passing through it;
$D_{50}$ and $D_{85}$ – diameter of particles for which 50% to 85% of the soil mass are finer;
$O_{50}$ – particle diameter for which 50% of the soil mass goes through the geotextile.

The PVDs available today present values of $q_w$ and $O_{90}$, which must be analyzed for each specific case, regarding permeability and granulometry (grain size) of the soil. The resistance and mechanic flexibility requirements of the filter and core are usually met.

PVDs that present a small reduction of $q_w$ when submitted to folding should be specified if very significant settlements or horizontal deformations are expected.

## 5.2.10 Sequence for radial drainage calculations

The definition of the spacing between drains in order achieve a given average rate of consolidation in soft clay deposit at a certain time has the following sequence:

1  Define the necessary geotechnical parameters: $c_v, c_h, k_v, k_h/k_h'$;
2  Define the installation layout, square or triangular pattern and relevant geometric values: $d_w, d_m, d_s$, and $h_{emb}$. The triangular pattern is more efficient and the square pattern is slightly easier to implement;
3  Estimate the discharge capacity of the PVD ($q_w$) for the representative state of stresses for the case;
4  Define the average degree of consolidation U to be achieved and the acceptable time ($t_{ac}$) to obtain U;

*Figure 5.8*  (A) horizontal drains; (B) detail of drainage well in (reinforced) embankment.

5  Define whether combined drainage will be considered or just radial drainage (more conservative);

6  Define spacing l (first attempt) and calculate $d_e$;

7  Calculate $T_v$ using Eq. (4.7) and, with Terzaghi's theory, calculate the corresponding $U_v$ (Figure 4.9), for the time $t_{ac}$ defined in step 4, if combined drainage is adopted;

8  Calculate F(n) using Eq. (5.8), which should be added to Eq. (5.17) to include the disturbance effect and to Eq. (5.21b) if the well resistance of the PVD is relevant;

9  Calculate $U_h$ using Eq. (5.16), with to $U_v$ calculated in step 7 for the assumed U value. If only radial drainage is considered $U_h = U$;

10  Use Eq. (5.6) to calculate the value of $T_h$ with the value of $U_h$ obtained in step 9 and F(n) in step 8; and using Eq. (5.7), the necessary time $t_{calc}$ to obtain the desired consolidation;

11  If $t_{calc} > t_{ac}$, gradually decrease l; and repeat steps 6 through 13, until $t_{calc} < t_{ac}$.

The typical spacing of PVDs usually varies between 2.5 m and 1.5 m, depending on the work schedule and the parameters of the compressible soil.

## 5.3  DESIGN OF THE HORIZONTAL DRAINAGE BLANKET

When using PVDs, the water flow reaching the base of the embankment is of such magnitude that a drainage blanket must be used and adequately designed so as not to delay the consolidation process. In this case, horizontal gravel drains wrapped in non-woven geotextile and known as "French drains" may be used inside the sand layer Figure 5.8A). They should also be used when pumping water from the drainage wells (Figure 5.8B) installed at the intersection of the French drains (Sandroni and Bedeschi, 2008).

Cedergren (1967) developed a method to calculate the total head loss $h_{cd}$ in a drainage blanket. As shown in Figure 5.9, for a square drain layout with spacing l, and for a conservative case of impermeable bottom layer, the discharge $q_d$ per drain is

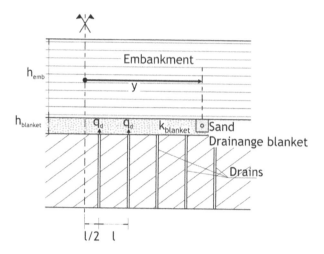

Figure 5.9 Detail of drainage mattress.

equal to the settlement rate r (equal $\Delta h/t$) times $l^2$ and y is the distance of the centerline of the embankment to the point of (French drain) interest, or:

$$q_d = r \cdot l^2 \tag{5.25}$$

where the value of r should be estimated from the settlement curve versus time for the start of consolidation. The height of load bearing loss in the drainage blanket is then defined using the equation:

$$h_{cd} = q_d \cdot y^2/(2 \cdot k_{blanket} \cdot A \cdot l) \tag{5.26}$$

where $k_{blanket}$ is the coefficient of permeability of the blanket material and A is the area of the blanket that refers to a line of drains. For drain spacing l and blanket with thickness $h_{blanket}$,

$$A = l \cdot h_{blanket} \tag{5.27}$$

Substituting (5.25) and (5.26) in (5.27)

$$h_{cd} = r \cdot y^2/(2 \cdot k_{blanket} \cdot h_{blanket}) \tag{5.28}$$

Allowing the height of load bearing loss on the drainage blanket to be, at the most, equal to the thickness of the blanket, $h_{blanket}$

$$y^2 = 2 \cdot k_{blanket} \cdot h_{blanket}^2/r \tag{5.29}$$

Assuming that y is the maximum distance at which a French drain should be installed within a drainage blanket and that:

- $r = 1.5 \times 10^{-7}$ m/s, referring to a 80 cm settlement in two months, as shown in Figure 5.9;
- $k_{blanket} = 10^{-4}$ m/s (lower limit for a coarse sand);
- $h_{blanket} = 0.50$ m for thickness of drainage blanket;

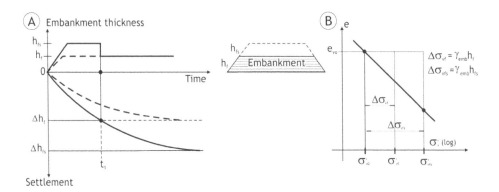

Figure 5.10 Acceleration of settlements with temporary surcharge.

Then $y = 18\,\mathrm{m}$, i.e. it is necessary to install French drains at a distance of $2y = 36\,\mathrm{m}$ from each other.

## 5.4  USE OF TEMPORARY SURCHARGE

Temporary surcharge has two main objectives: accelerate settlements through primary consolidation and compensate secondary compression settlements, in order to minimize post-construction settlements.

The use of temporary surcharge to accelerate settlements is shown in Figure 5.10, which shows a long term primary settlement $\Delta h_f$ for the applied vertical stress of $\Delta \sigma_{vf}$ (corresponding to embankment thickness $h_f$). A surcharge with full embankment height would cause a primary accumulated settlement to infinity equal to $\Delta h_{fs}$. When the surcharge is removed (thickness $h_s = h_{fs} - h_f$) at time $t_1$, the settlement stabilization time is accelerated. The removal of the surcharge may be followed by a slight rebound of settlements, which may not be measured in the field.

For a simplified case (disregarding submersion) of a normally consolidated clay layer, the settlements $\Delta h_f$ respectively $\Delta h_{fs}$ may be defined by:

$$\Delta h_f = \frac{h_{clay}}{1 + e_{vo}} C_c \log\left(\frac{\sigma'_{vo} + \Delta\sigma_{vf}}{\sigma'_{vo}}\right) = \frac{h_{clay}}{1 + e_{vo}} C_c \log\left(1 + \frac{\Delta\sigma_{vf}}{\sigma'_{vo}}\right) \tag{5.30}$$

$$\Delta h_{fs} = \frac{h_{clay}}{1 + e_{vo}} C_c \log\left(\frac{\sigma'_{vo} + \Delta\sigma_{vfs}}{\sigma'_{vo}}\right) = \frac{h_{clay}}{1 + e_{vo}} C_c \log\left(1 + \frac{\Delta\sigma_{vfs}}{\sigma'_{vo}}\right) \tag{5.31}$$

For the purpose of calculating the removal time $t_1$ for the surcharge, a consolidation rate of $U_s$ can be defined for the total applied stress equal to the total applied $\Delta\sigma_{vfs}$ by:

$$U_s = \frac{\Delta h_f}{\Delta h_{fs}} \tag{5.32}$$

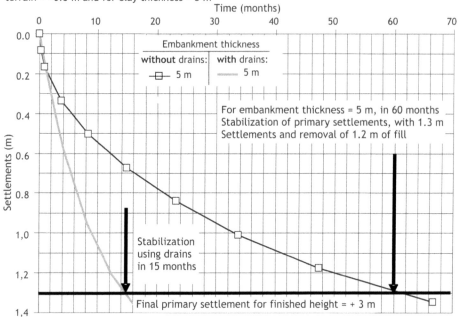

Figure 5.11  Use of surcharge with and without vertical drains.

Substituting Equations (5.30) and (5.31) in (5.32):

$$U_s = \frac{\log\left(1 + \frac{\Delta\sigma_{yf}}{\sigma'_{vo}}\right)}{\log\left(1 + \frac{\Delta\sigma_{yfs}}{\sigma'_{vo}}\right)} \tag{5.33}$$

## 5.4.1  Use of surcharge with and without vertical drains

Temporary surcharge may be associated with vertical drainage, radial drainage, or both. This section analyses these possibilities.

A case of a 6.0 m thick clay layer is used as an example. The top of the clay layer is at elevation +0.5 m, and the requirement is that primary consolidations settlements are stabilized for an embankment at elevation +3.0 m. Preliminary calculations indicate that final primary settlements in this case are $\Delta h_f = 1.3$ m, as shown in Figure 5.11.

To achieve the desired elevation, a fill embankment height $h_{emb}$ should be placed, equal to the difference between the original and final heights, plus the settlement value to be compensated, i.e. $h_{emb} = 3.0 - 0.5 + 1.3 = 3.8$ m. The surcharge thickness $h_s$ to be removed is equal to the difference between the two thicknesses of embankment, i.e. 1.2 m (=5.0 − 3.8 m).

Figure 5.11 shows settlement-time curves for situations of pure vertical drainage and combined drainage (radial plus vertical). It may be noted that the time to reach

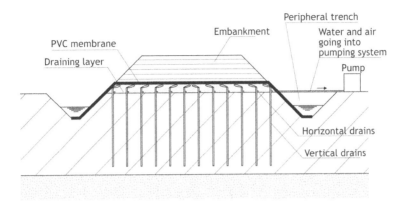

*Figure 5.12* Cross section scheme of vacuum preloading.

1.3 m without drains is 60 months, but the time to reach the same settlement with drains spaced 1.5 m apart is 15 months. The submersion of the embankment was considered in this analysis.

## 5.4.2 Vacuum preloading

Vacuum preloading (Kjellman, 1952; Chai, Bergado and Hino, 2010) is a special case of temporary surcharge, associated with vertical and horizontal drains. The vacuum is applied through a pumping system associated with the horizontal drains installed in the sand layer. There are two ways to apply vacuum (isotropic) pressure to the ground (Chai et al., 2008): in the first one, a waterproof PVC membrane covering the entire area and going down to the peripheral trenches prevents air from entering the system, thus creating the vacuum and ensuring that the system is watertight (Figure 5.12).

The pumping system, capable of pumping water and air simultaneously is coupled to a reservoir inside which the vacuum is almost perfect, with value of 100 kPa. However, the suction value measured under the membrane is approximately 70 to 75 kPa, which guarantees the system's efficiency around 70%–75%. When vacuum is applied, the soil's pore pressure may be reduced (Figure 5.13) until the final suction profile, at the end of the consolidation process. The pore pressure varies according to the position of the point in relation to the drain and time (u (radius, time)). The longer the pump is on, the higher the suction value inside the soil layer, possibly reaching a maximum of 70 to 75 kPa, i.e. the increase of effective stress of the soil corresponds to a surcharge equivalent to 4.5 m of embankment height.

Figure 5.13 presents a special case in which the groundwater level was 1.5 m below ground level which is represented by the initial hydrostatic profile $u_o$. When the pump system is turned on, the water level rises to the drainage layer and the hydrostatic profile is then $u_{ref}$. Thus, if the ground water table (G.W.L.) is deep, the vacuum system loses its efficiency in this case by 15 KPa. Therefore, at the end of consolidation, the variation of the effective stress will be the difference between the profiles, thus 60 kPa instead of 75 kPa. As a result, it is recommended to install the horizontal drains as close as possible to the G.W.L.

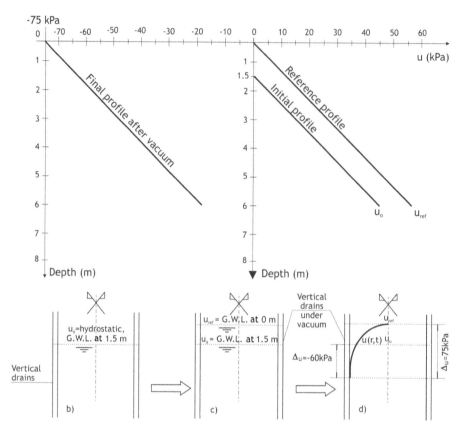

*Figure 5.13* Schematic pore pressures profile (adapted from Marques, 2001).

The second system of vacuum application does not use membrane, thus each PVD has a geosynthetic cap (it is sometimes called CPVD) and is connected to the pumping system individually, as shown in Figure 5.14. Chai et al., 2010 referred to this membraneless system as the vacuum-drain method, which may require more effort during the construction period. However, the versatility of application in different soil conditions, including submerged situations, makes the vacuum-drain method a suitable alternative to the conventional membrane system.

An advantage of the vacuum techniques over the conventional embankment loading is that there are no instability-induced failures since the stress path is always below the failure line due to the decreased pore pressure. Thus, this method does not require stabilizing lateral berms since there are no shear forces at the edges of the embankment, and vacuum preloading can be executed in a single step, thus speeding up the process.

After reaching the predicted settlements, the vacuum pumps are switched off and there is no need to dump material, thus minimizing the volume of earthworks. If an additional load is necessary, a surcharge may be used above the membrane, even during

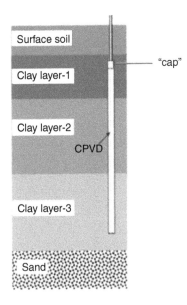

*Figure 5.14* Example of the vacuum-drain method (Chai et al., 2010).

the period in which the vacuum is applied, and this embankment can also be raised as the clay gains shear strength.

Practical difficulties occur when sand lenses are found in the soft clay to be treated, which may render the use of vacuum inadequate from a financial point of view. The execution of watertight walls down to the base of the sand layer can be a solution for improving the system efficiency (Varaksin, 2010). Furthermore, the vacuum pumps require electric installation, regular maintenance and safety measures from vandalism, which increases the costs of the technique. In the case of embankments in small areas, the technique may be less competitive because of high fixed costs.

### 5.4.3 Use of surcharge to minimize secondary compression settlements

It is possible to estimate values of $\Delta h_{sec}/h_{clay}$ according to the compression coefficient CR as shown in Chapter 4, Eq. (4.16). Thus, for highly compressible clay, with $CR = 0.50$, $\Delta h_{sec} \approx 7.5\% \cdot h_{clay}$, and for moderately compressible clay, with $CR = 0.25$, $\Delta h_{sec} \approx 3.8\% \cdot h_{clay}$. For these values and a clay layer of 10 m thickness for example, the settlement through secondary consolidation would vary between 75 cm and 38 cm.

The secondary compression behavior was observed experimentally by Garcia (1996), in samples collected in Barra da Tijuca (RJ), with CR values around 0.5 (Almeida et al., 2008c). Figure 5.15A shows compression curves for consolidation tests, where loading and unloading stages were performed to obtain an OCR line of approximately 2. It may be observed in Figure 5.15B that for effective vertical stress of 50 kPa, the vertical strain is approximately 7%.

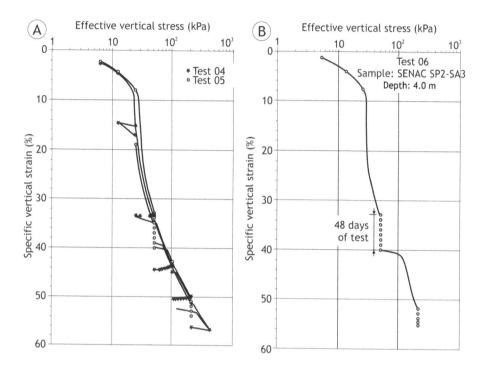

*Figure 5.15* Clay compression curve (adapted from Garcia, 1996).

The construction of a 2 m high embankment ($\Delta\sigma_v = 36\,kN/m^2$) on a 10 m thick clay layer, with specific weight 12.5 kN/m³ and water level at ground level will result in final vertical stresses $\sigma_{fv} = 50\,kPa$ and vertical strains due to secondary compression (see Figure 5.14B) around 7%. The conclusion is that the total settlement for secondary consolidation (very long term) may be an important portion of the construction's total settlements, this being as much important as primary consolidation settlements (Martins, Santa Maria and Lacerda, 1997).

It is then necessary to compensate these settlements so that they do not occur during the service life of the construction, which can be done with temporary surcharge. Usually, settlements caused by secondary consolidation are compensated during the construction period, i.e., even before the execution of paving works. Temporary surcharge is applied, followed by partial removal, so that the secondary compression settlements, calculated as proposed by Martins (2005), occur in the form of primary consolidation. Vertical drains are generally used to speed up settlements for thicker layers. In many cases, such surcharges must be applied in stages.

## 5.5   FINAL REMARKS

The use of PVDs as vertical draining elements, replacing the vertical sand drains, contribute to the improvement of the settlement stabilization technique of embankments on soft soils, particularly concerning execution speed and minimizing disturbances.

An important question to take into account when using surcharge for total compensation of secondary settlements is the high volume of earthworks required for clay deposits with high CR, low specific weight value and great thickness. For example, for a clay with specific submerged weight of $12\,kN/m^3$ (Barra da Tijuca clays) to generate an OCR of 1.5 and fully compensate for the secondary compression of the clay deposit of 10 m, an embankment thickness of about 3 m will be required. .

Due to the high compressibility and low strength of Brazilian clays, and because of the high values of secondary settlements, the use of PVDs with surcharge can become costly, due to high volumes of earthworks required, the need for strengthening and/or staged construction and the long construction periods. In such cases, the piled solution may be more suitable from an economic point of view, and also in terms of construction schedules.

# Chapter 6

# Stability of unreinforced and reinforced embankments

This chapter deals with the stability analysis of unreinforced and reinforced embankments constructed on soft clay deposits. Design parameters for the materials involved, clay foundation, embankment and geosynthetic reinforcement are discussed initially.

## 6.1 DESIGN PARAMETERS

### 6.1.1 Undrained strength of clay

Current stability analyzes consider undrained behavior of clay and are carried out based on total stress, due to their simplicity. The effective stress analysis is more complex (Bjerrum, 1972; Parry, 1972) as it requires the estimation of pore pressures generated in the soft clay layer.

The adopted undrained design strength $S_u$ for the clay layer is an essential information in total stresses, also known as analysis $\phi = 0$. The tests used to determine this were discussed in Chapter 2. Chart 6.1 summarizes the tests and procedures for defining the design strength $S_u$ to be used in stability calculations.

In general, the most widely used test for the determination of $S_u$ is the field vane test, to which a correction must be applied to obtain the strength to be used in the project, namely:

$$S_u \text{ (project)} = \mu \cdot S_u \text{ (Vane)} \tag{6.1}$$

The Bjerrum $\mu$ correction (1972, 1973) is the most used and results from the difference in shear strain rate of the Vane test compared to the shear strain rate of the embankment construction, in addition to the anisotropy effects of the clay. The Bjerrum $\mu$ values were obtained from back-analysis of embankment failures and are correlated with the plasticity index of the clay, as shown in Figure 6.1. This figure indicates data from failure analyses of some Brazilian cases, also showing the dashed curve proposed by Azzouz, Baligh and Ladd (1983), which relates to the correction to be used in the case of three-dimensional failures.

The CPTu is also used to obtain the undrained strength profile of the clay, with the advantage of defining a continuous $S_u$ profile, obtained by the equation:

$$S_u = \frac{q_t - \sigma_v}{N_{KT}} \tag{6.2}$$

*Chart 6.1* Procedures for measuring and estimating the undrained design strength $S_u$ (adapted from Leroueil and Rowe (2001), Duncan and Wright (2005) and the authors' experience).

| Tests / Procedures | Comments |
|---|---|
| Vane test | $S_u$ correction takes into account anisotropy effects and strain rate. The database used for the correction has reasonable dispersion. It is the most common procedure used due to its simplicity. The most applied correction factor is Bjerrum's (1972) (Figure 6.1), based on the plasticity index, but several others have also been proposed (Leroueil; Magnan; Tavenas, 1985). |
| Piezocone test | The empirical cone factor must be determined for the study area, correlating piezocone (CPTu) and Vane tests. In this case, the Bjerrum correction should be applied to the $S_u$ value. This procedure provides a continuous $S_u$ profile and soil layering. |
| UU triaxial test | The results tend to be more dispersed and they underestimate the strength due to sample disturbance; it should not be the single used procedure. |
| Triaxial and direct simple shear tests (DSS) | The triaxial compression and extension anisotropic tests CAU and direct simple shear tests (DSS) are conducted using recompression techniques (NGI) or SHANSEP. The disadvantages of these techniques are time and costs. The SHANSEP method is strictly applied to "mechanically overconsolidated" clays (Ladd, 1991) and tends to be conservative. |
| $\dfrac{S_u}{\sigma'_{vo}} = K \cdot OCR^m$ | Experimental based equation, where K and m parameters obtained through test program. For preliminary calculations, use $K = 0.23$ and $m = 0.8$ (Jamiolkowski et al., 1985) |
| $\dfrac{S_u}{\sigma'_{vo}} \cong \left(\dfrac{S_u}{\sigma'_v}\right)_{n.c.} \cdot OCR^\Lambda$ | Equation of Critical State Theory (Wood, 1990), where $\Lambda = 1 - C_s/C_c$ and $(S_u/\sigma'_v)_{n.c.}$ is the normalized strength in typically consolidated condition (see also Almeida, 1982) |
| $S_u = 0.22\sigma'_{vm}$ | The equation proposed by Mesri (1975) combines the influence of OCR and $\sigma'_{vo}$ in $\sigma'_{vm}$. Studies (e.g. Leroueil and Hight, 2003) indicate that the ratio $S_u/\sigma'_{vm}$ increases with the plasticity index, reaching values well above 0.22, especially for organic clays |
| OCR – overconsolidation ratio $(\sigma'_{vm}/\sigma'_{vo})$; $\sigma'_{vm}$ – overconsolidation stress; $\sigma'_{vo}$ – effective in situ vertical stress. | |

where the cone empirical factor $N_{kt}$ should be obtained from correlations between piezocone and vane tests ideally in the same deposit (Almeida et al., 2010). Figure 6.2 shows an example of $S_u$ profile of a deposit in the city of Rio de Janeiro, obtained from a piezocone test, which is compared with uncorrected Vane test data.

Special attention should be given to the $N_{kt}$ values used in Equation (6.2), since some authors report $N_{kt}$ values for uncorrected $S_u$ values, and others for already corrected values. The design $S_u$ values based on the piezocone tests should be corrected.

## 6.1.2 Embankment strength

The strength parameters of the embankment should be determined by laboratory testing. In general, the direct shear stress test is the most used. Tests on soils with natural moisture and on soil submersed close to saturation should be used to evaluate the variation of strength parameters in such conditions. It is usual to consider the saturated embankment with drained behavior, with $c = 0$ and $\phi \neq 0$ in the case of fill material with small amount of fines. However, in the case of large amount of fines a cohesion

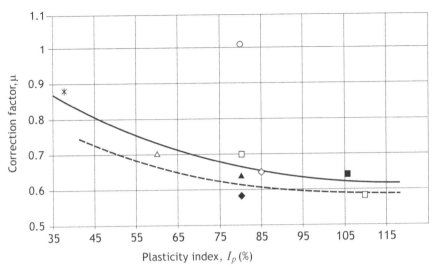

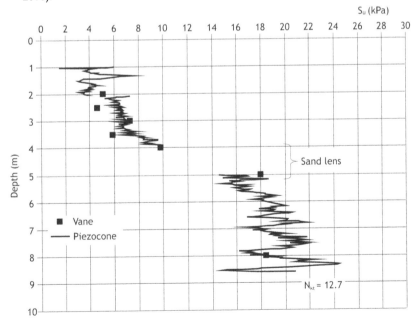

<table>
<tr><td>△ Massad (1999) Baixada Santista, SP</td><td>✳ Oliveira (2000) Recife, PE</td></tr>
<tr><td>□ Sandroni (1993) Duque de Caxias, RJ</td><td>◆ Magnani (2006) Florianópolis, SC</td></tr>
<tr><td>◇ Oliveira and Coutinho (2000) Recife, PE</td><td>▲ Magnani (2006) Florianópolis, SC</td></tr>
<tr><td>■ Almeida et al. (2008b) Recreio, RJ</td><td>— — Azzouz, Baligh and Ladd (1983)</td></tr>
<tr><td>○ Coutinho (1986) Juturnaíba, RJ</td><td>—— Bjerrum (1973)</td></tr>
</table>

*Figure 6.1*  Bjerrum correction factor (two-dimensional) and Azzouz (three-dimensional) applied to the vane test and back-analysis from Brazilian case histories (Almeida, Marques and Lima, 2010).

*Figure 6.2*  $S_u$ Profiles of CPTu and Vane tests ($S_u$ uncorrected).

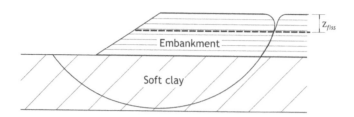

Figure 6.3 Depth of tension cracking in case of embankment material with fines.

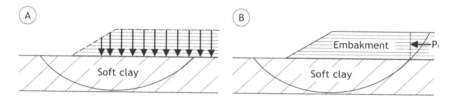

Figure 6.4 Stability analysis of cohesive embankments (A) totally cracked embankment, considered as surcharge, (B) lateral pressure in the case of embankments with low strength.

may be mobilized and the embankment may crack in its upper portion (Palmeira and Almeida, 1979), as shown in Figure 6.3.

The introduction of the tension crack also has the benefit of eliminating numerical instabilities in stability analyses, due to negative tensile stress (Duncan and Wright, 2005). The depth up to which the crack $z_{crack}$ develops is the one in which the horizontal stress is null, and is calculated by the equation:

$$z_{crack} = \frac{2 \cdot c_d}{\gamma_{(emb)} \cdot K_{a(emb)}t^{1/2}} \tag{6.3}$$

where:
$c_d$ – mobilized cohesion on the embankment;
$K_{a(emb)} = tg^2(45 - \phi_d/2)$ – active earth pressure coefficient of the embankment soil;
$\phi_d$ – mobilized friction angle on the embankment soil;
$\gamma_{(emb)}$ – Bulk specific weight of the embankment soil.

The embankment above the tension crack should be treated as a soil where $c = 0$ and $\phi = 0$, i.e., in this case the embankment can only be considered as a surcharge, as shown in Figure 6.4A. This consideration is not equivalent to the hypothesis of low values of c and $\phi$ because in this case an horizontal active force is installed, as shown in Figure 6.4B, resulting in different safety factors.

### 6.1.3   Geosynthetic reinforcement parameters

#### Types of geosynthetic reinforcement

Ehrlich and Becker (2010) have briefly showed the various types of geosynthetics used for soil reinforcement, as well as some relevant properties of these materials.

The two main types of geosynthetics used on embankments on soft soils are:

- Geogrids: Synthetic materials in a grid format specifically developed for soil reinforcement. They can be unidirectional when they present high tension strength in only one direction; or bidirectional, when they present high tension strength in both orthogonal directions.
- Geotextiles: Textile material that, due to the distribution of its fibers or filaments, can be woven, with filaments arranged in two orthogonal directions, or non-woven, with randomly arranged fibers.

The polymers used in geosynthetics also influence their performance as reinforcements. The most common polymers are polyester (PET), polypropylene (PP), polyethylene (PE) and polyvinyl alcohol (PVA).

It is acceptable to work with less resistant and less stiff materials, such as PET or PP geotextiles, with ultimate tensile strength typically between 30 and 80 kN/m for construction reinforcements on working platforms. However, materials such as woven PET or PVA geotextiles or geogrids, which have high stiffness modulus, high tensile strength and low susceptibility to creep, are recommended for structural reinforcements for embankments on soft soils. The typical nominal strength of these materials lies in the range between 200 and 1,000 kN/m, but there are geosynthetics with nominal strength of up to 1,600 kN/m being used in Brazil (Moormann and Jud, 2010).

### Tensile strength and stiffness modulus of the geosynthetic

By means of a tensile broadband test done on specimens that are 20 cm wide, one can obtain the load-strain curve of the geosynthetic, for a given loading condition. This is generally a non-linear curve and thus, one can calculate different stiffness moduli. It is customary to use the initial tangent modulus, which is the slope of the tangent line to the initial section of the curve, as well as the secant modulus, which is the slope of the line connecting the origin to a point on the curve – for example, 2% strain. The test provides the nominal tensile strength $(T_r)$, the nominal specific strain $(\varepsilon_r)$ and the nominal stiffness modulus $(J_r)$, which is the ratio between these two parameters. These values are often presented in manufacturer catalogs. However, they cannot be used directly in stability calculations, because the material suffers strength reductions at the worksite, mainly due to creep, in addition to installation damage and possible environmental degradation – chemical and biological.

Creep behavior of geosynthetics is determined by means of normalized tests, where specimens are subject to constant loads and strains are measured with time, until a given strain or failure occurs. The test is repeated for different load levels in order to obtain the failure loads due to creep and the isochronous load-strain curves for the desired loading times (one day, one month and one year, for example). One can interpolate and extrapolate for other load times based on the data from these curves. Figure 6.5A presents typical stress-strain curves for geogrids by the same manufacturer. One can observe the influence of the constituent polymer in the geogrid on its short-term stiffness and deformation, as well as the effect of the period of application of a constant load on the tensile strength and tension stiffness of a PVA geogrid (Figure 6.5B).

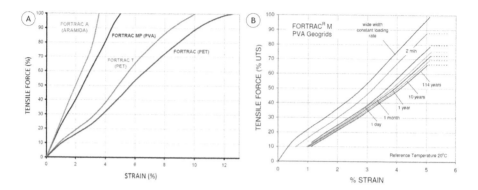

*Figure 6.5* (A) Accelerated Tensile Test – behavior of Fortrac geogrid made from different polymers, (B) Isochronous curves of Fortrac PVA geogrids obtained in creep tests (Source: Huesker).

### Tensile force T mobilized in reinforcement

The geosynthetic acts as a passive reinforcement. The foundation and embankment soils induce strains on the geosynthetic as they displace horizontally. The geosynthetic reacts and a resistant tensile force T is mobilized, restricting displacement of the soil layers.

The value of the tensile force T in the reinforcement to be used in stability calculations shall not exceed the tensile force limit that can induce $T_{lim}$, corresponding to the sum of the lateral pressures on the embankment and the shear force of the foundation soil. Thus:

$$T \leq T_{lim} = P_{emb} + P_{ref} \tag{6.4}$$

where:

$$P_{emb} = K_a(0.5 \cdot \gamma_{emb} h^2_{emb} + qh_{emb}) \tag{6.5}$$

$K_{a(emb)}$ is the active earth pressure coefficient of the embankment, calculated based on a reduced friction angle, according to:

$$\phi_d = tg^{-1}\left(\frac{tg\,\phi}{F_s}\right) \tag{6.6}$$

$$P_{ref} = X_T\left(\frac{\alpha S_{uo}}{F_s}\right) \tag{6.7}$$

where $S_{uo}$ is the undrained strength at the soil-embankment interface; $\alpha$ is the reduction factor applied to consider the reduction of the undrained strength at the compressible soil-embankment interface, and $X_T$ is the distance between the point where the circle intersects the reinforcement and the foot of the slope (Figure 6.6).

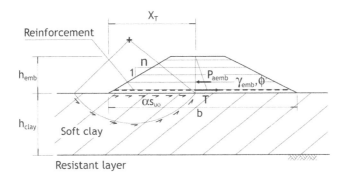

*Figure 6.6* Circular failure of an embankment on soft soil.

### Allowable strain in the reinforcement

The embankment height at failure and the T value calculated by means of the limit equilibrium methods do not guarantee proper behavior under working conditions. In some cases, embankments have failed due to excessive strains (serviceability limit state) before reaching the failure height (ultimate limit state). This has been recognized by several authors (e.g. Rowe and Sodemann, 1985; Bonaparte and Christopher, 1987), who recommend allowable strain values ($\varepsilon_a$) in the reinforcement in the range of 2% to 6%. Additionally, the British standard BS 8006 (BSI, 2010) recommend that the reinforcement should have a maximum strain of 5% for short-term applications, and between 5% and 10% for long-term conditions, where, in the case of sensitive soils, it should be even lower (<3%), to ensure strain compatibility with the foundation soil.

Rowe and Sodermann (1985) proposed a method applicable for foundations with constant strength and limited depth, and for embankments without berms. Using this method it is possible to evaluate the tensile force mobilized at the reinforcement from the strain value using a dimensionless parameter $\Omega$, defined in Equation (6.8). The allowable strain ($\varepsilon_a$) is defined as the maximum strain before the collapse of the embankment and, therefore, refers to a condition of unitary safety factor. The authors have defined the dimensionless parameter $\Omega$, which is related to $\varepsilon_a$ by means of the curve shown in Figure 6.7 based on extensive numerical studies on reinforced embankments on soft soils:

$$\Omega = \frac{\gamma_{emb} h_{cr}}{S_u} \frac{S_u}{E_u} \left( \frac{h_{clay}}{B} \right)^2 \tag{6.8}$$

where:
$h_{cr}$ – collapse height of the unreinforced embankment (see item 6.3);
B – width of the platform;
$h_{clay}$ – thickness of the soft layer;
$S_u/E_u$ – ratio between strength and undrained Young's modulus;
$\gamma_{emb}$ – specific weight of embankment material.

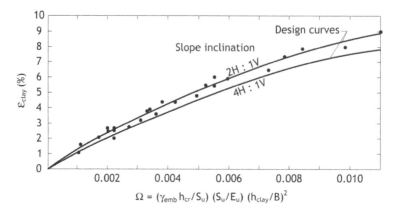

$$\Omega = (\gamma_{emb} \, h_{cr}/S_u) \, (S_u/E_u) \, (h_{clay}/B)^2$$

• Results of finite elements analyses for slope: 2h:1v

$(h_{clay}/B) = 0,2$   for $h_{clay}/B < 0,2$          $(h_{clay}/B) = 0,84 - h_{clay}/B$ for $0,42 < h_{clay}/B \leq 0,84$

$(h_{clay}/B) = h_{clay}/B$ for $0,2 \leq h_{clay}/B \leq 0,42$          $(h_{clay}/B) = 0$          for $0,84 < h_{clay}/B$

*Figure 6.7* Allowable strain according to geometric and geotechnical parameters.

The stress value in the reinforcement is calculated from the $\varepsilon_a$ value, as:

$$T = J \cdot \varepsilon_a \qquad (6.9)$$

where J is the stiffness modulus of the reinforcement.

From the definition $(h_{clay}/B)$ based on the $\varepsilon_a$ versus $\Omega$ curve, it is observed that in Rowe and Sodermann's model (1985), for values $(h_{clay}/B) > 0.84$, i.e., deep deposits, the reinforcement does not have a stabilizing effect for deep surfaces, although it improves the stability near the foot of the slope.

For the cases where the undrained strength increases with depth, Hinchberger and Rowe (2003) proposed similar abaci to the ones in Figure 6.7 to estimate $\varepsilon_a$.

### Anchoring the reinforcement

The geosynthetic must be appropriately anchored in the soil to induce the tensile force T. The length of the anchor ($L_{anc}$) is dictated by the strength parameters of the soil and soil-reinforcement interface. It can be calculated by:

$$L_{anc} = \frac{T_{anc}}{2 \times C_i \times (c + \gamma \times h \times tg \, \phi)} \qquad (6.10)$$

where:
$T_{anc}$ – anchoring strength ($T_{anc} \geq T$);
$C_i$ – geosynthetic-soil interaction coefficient, obtained through pullout tests;
$h$ – embankment height above reinforcement;
$\gamma, c, \phi$ – embankment soil parameters.

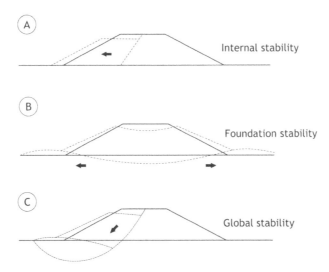

*Figure 6.8* Failure modes of unreinforced embankments: (A) lateral sliding of the embankment, (B) failure in foundation clay (Jewell, 1982), (C) overall failure embankment-foundation.

The $C_i$ values must be supplied by manufacturers. They may vary according to the type of geosynthetic. Geogrids with square mesh opening between 20 mm and 40 mm may have interaction coefficients $C_i$ greater than 0.8. For geogrids with larger openings and few transversal strips, $C_i$ may be less than 0.5. In the case of woven geotextiles, $C_i$ may be around 0.6.

## 6.2  FAILURE MODES OF EMBANKMENTS ON SOFT SOILS

Some possible failure modes of embankments on soft soils valid for unreinforced and reinforced embankments are shown in Figure 6.8. They include failure within the body of the embankment without involving the soft clay (Figure 6.8A); failure of foundation clay due to low bearing capacity (Figure 6.8B), and overall embankment-foundation failure (Figure 6.8C). The analysis of lateral extrusion of the soft soil (Palmeira and Ortigão, 2004) should also be verified. Strictly speaking, one should analyze all failure modes, but in general, the failure modes that govern the problem of an embankment on soft soil are those of foundation instability and overall failure, for which the analysis methods are discussed in the next sections. Details of the failure modes of reinforced embankments for ultimate limit state conditions and working limit state are given in BS 8006 (BSI, 1995).

## 6.3  FOUNDATION FAILURE: CRITICAL HEIGHT OF EMBANKMENT

Failure of the foundation of the embankment is a bearing capacity issue. In this case, for stability purposes, the embankment participates with load, but not with strength.

For stability analysis, abaci may be used to calculate the critical height $h_{cr}$ of the embankment on soft soil, and this is the first stage of the analysis. The equation used is derived from the classic bearing capacity equation of a direct foundation on $\phi = 0$ soil with undrained strength $S_u$, given by:

$$h_{cr} = \frac{N_c \cdot S_u}{\gamma_{emb}} \tag{6.11}$$

where $N_c$ is the bearing capacity factor. $N_c$ values for finite clay layer and gradual of $S_u$ profile increasing with depth are discussed in section 6.5.2.

The abaci developed by Pinto (1966) for strength gradual increase with depth are also easy to use. For construction purposes, if the inclination of the slope is small, it can be replaced with equivalent lateral berms (Massad, 2003).

The admissible height $h_{adm}$ adopted for the design of an embankment built in one stage is equal to:

$$h_{adm} = \frac{h_{cr}}{F_s} = \frac{N_c \cdot S_u}{\gamma_{emb} F_s} \tag{6.12}$$

where $F_s$ is the safety factor defined according to design criteria, considering the importance of the job. Generally, $F_s$ values greater than 1.5 are used, and lower values are acceptable ($F_s \geq 1.3$) for stability analysis of a temporary condition (e.g. staged embankments), with inclinometer monitoring and the absence of close neighbors.

If the value of $h_{adm}$ is less than the required embankment height $h_{emb}$ for the project, one must use an alternative construction method, such as, for example, staged construction or reinforced embankment.

## 6.4 GLOBAL STABILITY ANALYSIS OF UNREINFORCED EMBANKMENTS

### 6.4.1 Circular failure surfaces

One can adopt abaci for preliminary studies of global stability analysis of the embankment. Pilot and Moreau (1973) developed abaci for purely granular embankments with different slope inclinations, foundation with constant strength and circular failure surface. However, with the various slope stability programs available on the market, abaci are increasingly unpopular. The advantage of the abaci, however, is that the user can easily perceive the variations of $F_s$ due to the variables involved in the problem.

The method of slices are commonly used for stability analysis of embankments on soft soils, but there is no guarantee that such method provides the lowest $F_s$ value. Duncan and Wright (2005) have compared a number of stability analysis methods for circular failure of a purely granular embankment on soft soil (constant $S_u$), and the results are summarized in Table 6.1. The modified Bishop method has been widely used in geotechnical practice, but it does not necessarily provide the lowest $F_s$. The calculation of the safety factor in this case, by the method of wedges or blocks (described in section 6.4.2), resulted in $F_s = 1.02$, about 16% lower than the value presented by Bishop's method.

*Table 6.1* Comparison between results of methods of slices for circular surfaces of granular embankment on soft soil (adapted from Duncan and Wright, 2005).

| Method of slices | Safety factors |
|---|---|
| Fellenius | 1.08 |
| Bishop | 1.22 |
| Spencer | 1.19 |
| Simplified Janbu with correction | 1.16 |
| Simplified Janbu without correction | 1.07 |

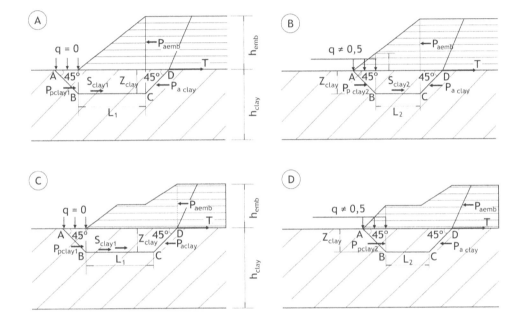

*Figure 6.9* Wedge method for planar surfaces: (A) failure outside the foot of the embankment, without berms, (B) failure at the foot of embankment, without berms, (C) failure outside the foot of the embankment, with berms, (D) failure at the foot of embankment, with berms.

## 6.4.2   Non-circular failure surfaces

Non-circular failure surfaces must also be analyzed, and the simplified Janbu method (Janbu, 1973) is frequently used for this. These surfaces must also be analyzed by the method of wedges or blocks, also called translational analysis, easily developed in spreadsheets or computer programs. A typical scheme for stability analysis by this method is shown schematically in Figure 6.9A. In this method, the safety factor is the result of dividing the sum of resisting forces by the sum of the unstabilizing forces, according to the equation:

$$F_s = \frac{P_{p\,clay} + S_{clay}}{P_{a\,clay} + P_{emb}} \tag{6.13}$$

where:

i.  $P_{p\,clay}$ is the passive force in the clay, equal to:

$$P_{p\,clay} = \frac{1}{2}\gamma_{clay} \cdot z_{clay}^2 \cdot K_{p\,clay} + 2S_u \cdot z_{clay}\sqrt{K_{p\,clay}} + q \cdot z_{clay} \cdot K_{p\,clay} \qquad (6.14)$$

  where q is the vertical stress acting on the top of the clay layer, with $q = 0$ for the case of Figure 6.9A ($P_{pclay1}$) and $q \neq 0$ for the case of Figure 6.9B ($P_{pclay2}$);

ii.  $S_{clay}$ is the mobilized shear force in the soft clay, equal to:

$$S_{clay} = S_u L \qquad (6.15)$$

  where L is the horizontal distance of the failure line traversing the clay to a depth $z_{clay}$ (see $L_1$ and $L_2$ in Figure 6.9) and $S_u$ is the undrained strength of the clay at this depth;

iii.  $P_{aemb}$ is the active force in the sandy embankment, not considering cohesion, and equal to:

$$P_{aemb} = \frac{1}{2}\gamma_{emb} \cdot h_{emb}^2 \cdot K_{aemb} \qquad (6.16)$$

iv.  $P_{a\,clay}$ is the active force in the clay layer, equal to:

$$P_{a\,clay} = \frac{1}{2}\gamma_{clay} \cdot z_{clay}^2 \cdot K_{a\,clay} - 2S_u \cdot z_{clay}\sqrt{K_{a\,clay}} + \gamma_{emb} \cdot h_{at} \cdot z_{clay} \cdot K_{a\,clay}$$
$$(6.17)$$

• The following observations can be made for analysis type $\phi = 0$, $K_{a\,clay} = K_{p\,clay} = 1$ in Equations (6.14) and (6.17).
• The equations presented should be adapted to the case of sand layers in the embankment foundation.

One must evaluate safety issues, considering failures at various depths within the clay layer, and thus obtaining different $F_s$ values according to depth.

In cases of low strength layers, or long berms, safety factors calculated by wedge methods and non-circular surfaces tend to be lower than when they are calculated using circular surfaces. Duncan and Wright (2005) reported the case of James Bay dyke, approximately 4 m high, built over clay layers with different $S_u$ values. Calculation using circular surfaces resulted in $F_s = 1.45$. However, the calculation using non-circular surfaces resulted in $F_s = 1.17$ (this value coincides with the wedge method), i.e. a difference of approximately 20%.

The wedge method has the advantage of guaranteeing complete control of calculations and the various components of the $F_s$ equation. It is particularly useful for reinforced embankments as in this case the calculation hypotheses adopted in some programs, are not always available to the user. Therefore, it is recommended, that these computational analyses be also verified by the wedge method, as outlined above.

## 6.5   REINFORCED EMBANKMENTS

### 6.5.1   Effects of reinforcement

The earth pressure that develops inside an embankment causes outward shear stresses (Figure 6.10A – without reinforcement), similar to the behavior of a smooth footing (Figure 6.10B). These shear stresses reduce the load bearing capacity of the clay foundation (Figure 6.10D). The reinforcement placed on the base of the embankment has two functions: resist earth pressures developed inside the embankment (Figure 6.10A – with reinforcement) and resist the lateral deformation of the foundation, thus changing the shear stress direction (Figure 6.10C). The latter is similar to the behavior of a rough footing. The reinforcement increases the load bearing capacity of the foundation as shown in Figure 6.10E (Leroueil and Rowe, 2001). As a result, reinforced

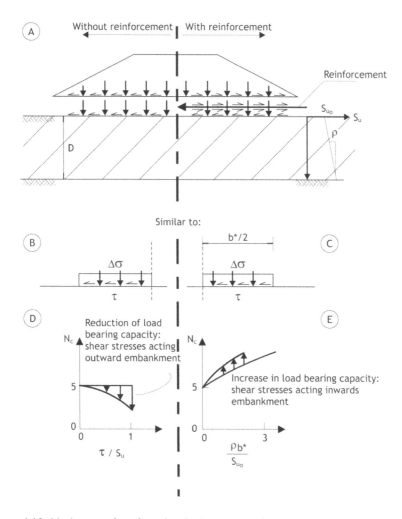

*Figure 6.10*  Mechanism of reinforced embankment on soft clay (Leroueil and Rowe, 2001).

embankments can reach greater heights than unreinforced embankments, or, comparing a unreinforced embankment with a reinforced one of the same height, there is a gain of $F_s$ with the reinforcement.

The failure modes of reinforced embankments – in essence the same as unreinforced embankments shown in Figure 6.8 – are analyzed separately below.

### 6.5.2   Foundation failure

As with unreinforced embankments, the preliminary step when analyzing the stability of reinforced embankments consists of checking if the foundation has the bearing capacity to withstand stresses on the reinforced embankment. Admittedly, this acts as a rough footing on the clay layer, as shown in Figure 6.10C. In this case, the rough footing simulates the insertion of the reinforcement on the embankment and the abaci by Soderman and Rowe (1985) can be used for $S_u$ increasing with depth (Figure 6.11A); or for $S_u$ constant with depth; (Mandel and Salençon, 1972; Davis and Booker, 1973) (Figure 6.11B).

This preliminary step makes it possible to define the maximum embankment height $h_{emb}$ to be used. It is recommended that the bearing capacity required from the unreinforced embankment be lower than the admissible stress of the clay, so that the reinforcement is not fully responsible for the stability, i.e., the factor of safety of the unreinforced embankment is at least equal to one.

### 6.5.3   Failure due to lateral sliding of embankment

One step is the analysis of the failure due to sliding of the sandy embankment at its base (above the reinforcement), due to earth pressure of the embankment. When the equilibrium of the horizontal forces is determined, as shown in Figure 6.12, the safety factor is given by:

$$F_s = \frac{0.5 \cdot n \cdot \gamma_{emb} h_{emb}^2 \, tg\, \phi_d}{K_a(0.5 \cdot \gamma_{emb} h_{emb}^2 + q h_{emb})} = \frac{0.5 \cdot n \cdot \gamma_{emb} h_{emb} \, tg\, \phi_d}{K_a(0.5 \cdot \gamma_{emb} h_{emb} + q)} \tag{6.18}$$

where n is the inclination of the slope; $K_{aemb}$ is the active earth pressure coefficient, and $\phi_d$ is the reinforcement-soil friction.

The failure safety factor due to sliding below the reinforcement can be calculated by the equation:

$$F_s = \frac{n \cdot S_{clay} + T}{K_a(0.5 \cdot \gamma_{emb} \cdot h_{emb} + q)} \tag{6.19}$$

where $S_{clay}$ is the mobilized force in the reinforcement-clay contact and T is the induced tensile force in the reinforcement.

Christopher, Holtz and Berg (2000) recommend $F_s \geq 1.5$ for the lateral sliding analyses.

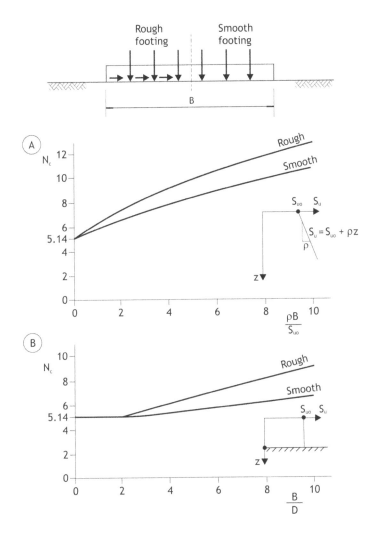

Figure 6.11  Load bearing capacity factor smooth and rough footing: (A) $S_u$ increasing with depth; (B) $S_u$ constant.

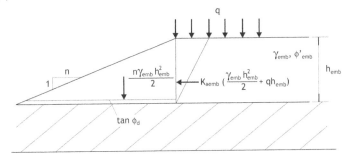

Figure 6.12  Sliding of the embankment.

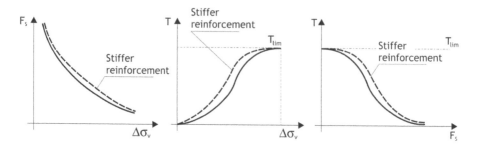

*Figure 6.13* Behavior of reinforced embankments on soft soils (Magnani, Almeida and Ehrlich, 2009).

### 6.5.4 Global failure

One can adopt abaci for preliminary studies for global stability analysis of the reinforced embankment and its foundation. The method of Low, Wong; Lim (1990) for circular surface failure, can be useful in preliminary analyses.

Some computer programs for stability calculations consider the contribution of the reinforcement (Figure 6.9) in the numerator of the safety factor equation, as in Eq. (6.13) for the wedge method increasing the strength (or the resisting moment in the case of slice methods) i.e., by:

$$F_S = \frac{P_{p\,clay} + T + S_{clay}}{P_{a\,clay} + P_{emb}} \tag{6.20}$$

Other programs consider that the reinforcement decreases the active stresses (the contribution of the reinforcement appears in the denominator of the safety factor equation). Therefore, computer programs currently available must first be verified before use (Duncan and Wright, 2005) because the safety factors results are different in each case.

### 6.5.5 Definition of tensile force in reinforcement

*Relationship between safety factor $F_s$, vertical stress of embankment $\Delta\sigma_v$ and tensile force T in reinforcement*

Figure 6.13 show schematically the relationship between $F_s$ and the stress applied by the embankment ($\Delta\sigma_v$) and the induced tensile force in the reinforcement. The limit stress ($T_{lim}$) presented in this figure refers to the completely yielded foundation (soft soil). The influence of reinforcement stiffness (J) is schematically presented, indicating that, for the same load, a stiffer reinforcement will induce a higher T value; therefore, a higher safety factor will be achieved and, for the same T value, a reinforcement with greater stiffness will result in increased $F_s$.

Table 6.2 Range of reduction factor values to be used in Eq. (6.21).

| Reduction factors | Geotextile | Geogrid |
|---|---|---|
| $FR_I$ | 1.1–2.0 | 1.1–2.0 |
| $FR_{DQ}$ | 1.0–1.5 | 1.1–1.4 |
| $FR_{DB}$ | 1.0–1.3 | 1.0–1.2 |
| $FR_F$ | 2.0–3.5 | 2.0–3.0 |

### Specification of the reinforcement to be used

The mobilized tensile force T, the allowable strain $\varepsilon_a$ and the coefficient of interaction $C_i$ are design parameters used in stability analysis. Thus they should be included in the specification of the reinforcement to be used.

The force of the geosynthetic T calculated should be compared with the admissible tensile strength ($T_{adm}$) of the material ($T_{adm} \geq T$). The admissible strength ($T_{adm}$) can be calculated from the nominal tensile strength ($T_r$) obtained in the broadband test, according to the equation:

$$T_{adm} = \frac{T_r}{FR_F \times FR_I \times FR_{DQ} \times FR_{DB}} \tag{6.21}$$

where:
$FR_F$ – partial reduction factor due to the creep for the project service life;
$FR_I$ – partial reduction factor due to mechanical damage during installation;
$FR_{DQ}$ – partial reduction factor due to chemical degradation;
$FR_{DB}$ – partial reduction factor due to biological degradation.

The recommended reduction factor values (Koerner and Hsuan, 2001) for general embankment design are presented in Table 6.2 and should be used based on experience and judgment.

One can determine the stiffness modulus J from Equation (6.9) by applying the tensile force T and allowable strain $\varepsilon_a$, thus using the equation:

$$J = \frac{T}{\varepsilon_a} \tag{6.22}$$

It is necessary to increase the design stiffness modulus J to define the nominal stiffness modulus $J_r$, by considering creep effect for the service life of the construction, which can be obtained by means of isochronous curves. Damages caused during installation, chemical and biological degradation factors should also be applied to this value, as shown in Table 6.2. In the absence of isochronous curves, one can adopt an approximate method, by multiplying J by all partial reduction factors, including creep reduction factor, to determine $J_r$ (nominal) to be specified.

The specification of a reinforcement design through its stiffness modulus has the benefit of associating a given tensile strengths to a given strain, leading to better choices when selecting geosynthetics for the construction.

## 6.6   STABILITY ANALYSIS OF STAGE CONSTRUCTED EMBANKMENTS

### 6.6.1   Conceptual aspects

The effective stress path of a clay element, located below the center of an embankment constructed by stages is shown schematically in Figure 6.14 (Leroueil and Rowe, 2001) for a lightly overconsolidated clay. The in situ stress state $I_0$, changes during the first construction stages, to $C_1$ and $E_1$ (along the limit state curve).

During the consolidation stage of the first loading step the state of effective stresses moves from $E_1$ to $E'_1$, and the effective stress path moves away from the failure envelope, thus there is an increase in the clay strength, and consequently in $F_s$. When the embankment height is raised at the end of stage one, the effective stress path goes from $E'_1$ to $E_2$, i.e. the effective stress path moves towards the failure envelope. For stage three, starting at the end of stage two (point $E'_2$), the embankment was raised until the failure point in R. It is therefore necessary to evaluate the $F_s$ for each loading stage, in regards to the increase of the undrained strength $S_u$ of the clay, which occurs along the effective stress path shown in Figure 6.14.

The Leroueil, Magnan and Tavenas (1985) schematic model is illustrated in Figure 6.15 by means of numerical analyses (Almeida, Britto and Parry, 1986) using the modified Cam-clay (isotropic yield locus) model for various points of the clay layer schem presented at the bottom right side of Figure 6.15.

### 6.6.2   Undrained strength of the clay for staged construction

Stability analyses on staged embankments are commonly carried out in terms of total stress, by estimating the undrained strength of the clay foundation layer prior to placing the next fill layer, i.e. for the stress states indicated as $E'_1$ and $E'_2$ in Figure 6.14.

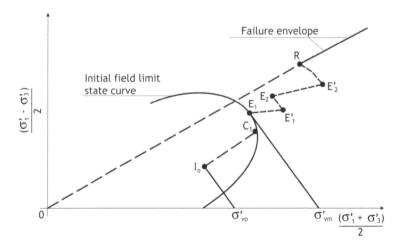

*Figure 6.14* Schematic stress path of a clay element located below the center of the embankment (Leroueil, Magnan and Tavenas 1985).

Chart 6.2 presents the more widely used methods for estimating clay strength, in which $\sigma'_{v1}$ (see Eq. 4.19) is the effective stress at the respective depth, due to the embankment loading during that stage. If $\sigma'_{v1} < \sigma'_{vm}$, one must adopt $\sigma'_{v1} = \sigma'_{vm}$. When evaluating the effective stress at the end of the first stage, one must consider the submersion effect of the embankment, which is also considered in the second part of the equation. This procedure is similar for all subsequent stages. Furthermore, for the evaluation of the gain in strength $\Delta S_u$, one must consider the settlements, thus the clay depth variation, as seen in Figure 6.16 and using the normalized depth as shown in Table 6.3.

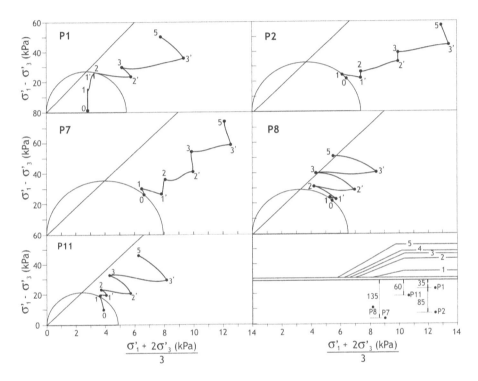

*Figure 6.15* Stress path for staged constructions: numerical modeling (Almeida, Britto and Parry, 1986).

*Chart 6.2* Procedures for estimating the clay undrained strength $S_u$ for stability calculations of staged constructed embankments

| Tests/Procedures | Comments |
| --- | --- |
| Variation of the effective stresses $\sigma'_v$ in the clay layer | The equation $S_u = 0.25 . \sigma'_v$ (Leroueil, Magnan and Tavenas, 1985, Wood, 1990) is analogous to Mesri (1975) equation $S_u = 0.25 \cdot \sigma'_{vm}$, and has been proven to be valid (e.g. Almeida et al., 2001). The relationship $S_u/\sigma'_v$ can be obtained by CAU triaxial tests |
| Vane tests before the next construction stage | It is recommended to measure the clay strength $S_u$ in order to evaluate the gain in clay strength adopted in the design phase. Studies indicate that the Bjerrum correction should not be applied in this case (Leroueil et al., 1978; Law, 1985) |

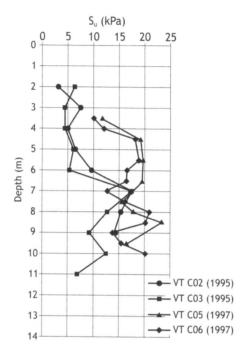

*Figure 6.16* Vane tests: before and after construction of the embankment – soft clay layer with PVDs (Almeida et al., 2001).

*Table 6.3* Strength gain in clay after the construction of embankment.

| Normalized soft clay Depth d (m) | $\Delta S_u/\Delta\sigma_v'$ |
| --- | --- |
| 0.35 | 0.25 |
| 0.40 | 0.34 |
| 0.45 | 0.47 |
| 0.50 | 0.46 |
| 0.55 | 0.32 |
| 0.60 | 0.06 |
| 0.65 | 0.05 |
| 0.70 | 0.05 |
| 0.75 | 0.17 |

### 6.6.3   Illustration of stability analysis of staged construction

The stability calculation for the next stage should be done with the new strength profile calculated as explained in the previous section and in Table 6.2. For staged constructed embankments, one should take into account the new geometry of the problem, considering the decrease of the soft layer thickness and the submersion of the embankment.

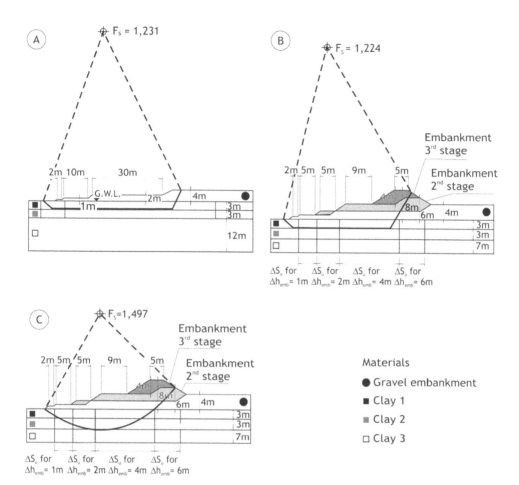

*Figure 6.17* Stability analysis of a dam in the port area: (A) 1st stage of construction ($h_{emb} = 4\,m$), (B) 3rd stage of construction ($h_{emb} = 8\,m$), non-circular failure, (C) 3rd stage of construction ($h_{emb} = 8\,m$), circular failure.

Figure 6.17 presents the results of stability analyses for staged construction of a reinforced dyke built on vertical drains with berms. In the analyzed soft soil deposit, the strength profile increased with depth, intercepting the origin close to zero and increasing with at a rate of approximately 1.2 kPa/m. The analyses were performed considering the clay strength gain at each load stage. Furthermore, the gradual increase of the induced tension in the reinforcement with the stages was also considered in the analyses.

The results of stability analyzes presented in Figure 6.17 are summarized in Table 6.4. Factors of safety arising from non-circular failures are substantially lower than from circular surfaces, which confirm other results mentioned before. The values of $F_s$ are those at the end of the loading, which are extremely low (points $E_1$ and $E_2$ in Figure 6.14). For embankments on soft soils in areas without nearby construction $F_s$ values higher than 1.3 may be accepted provided inclinometer monitoring is used to assess performance (see Chapter 8), but $F_s$ values in the order of 1.2 are indeed quite low.

Table 6.4 Safety factor values in circular and non-circular failure analyses.

| Stage | Thickness of the embankment (m) | Safety factors | |
|---|---|---|---|
| | | Non-circular failure | Circular failure |
| 1st | 4 | 1.23 | 1.81 |
| 2nd | 6 | 1.29 | 1.56 |
| 3rd | 8 | 1.22 | 1.50 |

### 6.6.4   Considerations on the stability analysis for staged constructed embankments

The in situ vane test is the recommended tool to obtain the clay undrained strength prior to the construction of the next stage, so as to evaluate if the resistance considered in the design can actually be verified.

For reinforced embankments, the contribution of the reinforcement should be considered in the new stage of the embankment; however, it is important to evaluate the new geometry for this type of situation as well, including considerations of the creep effect on reinforcement strength, in case there is a significant time interval between the stages.

The use of PVDs speeds up the consolidation process; thus accelerating the strength gain of the clay. It is recommended that PVDs be installed up to at least half the length of the embankment slope or up to half the length of the equilibrium berm (see Figure 1.5A), in the case of staged embankments. This procedure contributes to the quick increase in clay strength under this area, which may be considered in stability analyses of the next stages. In the case of very soft clays it is also recommended to correct the settled berm height before the implementation of the second and subsequent stages.

### 6.7   SEQUENCE FOR STABILITY ANALYSIS OF EMBANKMENTS ON SOFT SOILS

One should calculate the admissible height of the embankment (Equation 6.12) once the dimensioning parameters and the safety factor to be adopted are defined. The definition of the admissible height serves as preliminary. In this phase, it is already possible to decide whether the embankment should be built in stages or reinforced, or if both solutions are to be adopted.

### 6.7.1   Unreinforced embankments

The proposed sequence to assess the stability of unreinforced embankments is:

1   Stability assessment of embankment with allowable height considering circular and non-circular failure surfaces, as discussed in section 6.4, by analyzing the stability at different slope inclinations or the use of equilibrium berms.

    a.   If the $F_s$ obtained is satisfactory there is no need to use reinforcements or staged construction.

b.  If the $F_s$ obtained is lower than required, one should evaluate the stability for staged construction or the use of reinforced embankment or both.

2  If the adopted solution is for a staged embankment without reinforcement, one should predefine the duration of the stages and the thickness of the embankment for each stage, depending on the available deadlines.

3  Stability calculations for embankment elevation in each stage should be carried out, considering circular and non-circular surfaces. One should also consider the strength gain for the next steps, as well as the geometry change of the problem, since with the settlements there is a decrease in the soft layer thickness and submersion of the embankment.

In the case of very soft soil the use of geosynthetic reinforcements at the base of the embankment is almost mandatory.

## 6.7.2  Reinforced embankments

The proposed sequence to assess the stability of reinforced embankments is:

1  Stability assessment of unreinforced embankment with admissible height considering circular and non-circular failure surfaces, as previously discussed. If $F_s$ is lower than in the design, the choice then should be for a reinforcement solution at the base of the embankment to increase $F_s$.

2  Define the T value.

a.  Consider the lateral sliding of the embankment – Equation (6.19).
b.  Consider wedge failure – Equation (6.20).
c.  Or by means of circular failures, which can be assessed using available stability programs.

The highest T value between the values calculated in (2.a), (2.b) and (2.c) is adopted for $F_s$ design, and that value must meet the $T_{lim}$ criteria discussed in section 6.1.3. If not so, the geometry of the problem should be changed (height or slope of the embankment) and the calculations above should be repeated.

1  Definition of allowable strain $\varepsilon_a$ and stiffness modulus:

a.  Adopt the $\varepsilon_a$ value based on local experience, considering the points presented in section 6.1.3. For foundation soils with constant resistance and limited depth, use equation (6.8) and the abacus in Figure 6.7 to determine the admissible strain $\varepsilon_a$ on the geosynthetic.
b.  Use T and $\varepsilon_a$, in equation (6.22) to calculate the stiffness modulus J.

2  Verification of anchoring length:
Use Equation (6.10) to determine if the anchoring length is sufficient to mobilize the tensile force T in the reinforcement. This verification should be done for active and passive areas of failure wedges.

3  Specification of geosynthetic reinforcement:
Once T and J are defined as design parameters, it is necessary to specify the nominal properties of these materials, to allow their acquisition and verification at

the worksite. Consider the reduction factors presented in Equation (6.21) to define the reinforcement to be used. Compare the calculated T in (2) with $T_{adm}$ obtained from Equation (6.21), for the choice of reinforcement to be used, considering reinforcement service life, which could be the reinforcement acting only during construction and consolidation stages and/or later. Similarly, the minimum value of the stiffness $J_R$ should be calculated.

### 6.7.3   Reinforced embankment built in stages

For the condition described in 6.7.1 (1b) and after defining the design parameters and the safety factor to be adopted, one should perform the preliminary calculation of the admissible height of the embankment (Eq. 6.12) using $N_c$ values for rough interface conditions shown in Figure 6.11. If the admissible height is lower than the required height, a reinforced embankment should be built in stages. In general, one has to evaluate the cost effectiveness of using reinforcements with higher $T_r$ and $J_R$ values to minimize the number of construction stages. The use of equilibrium berms can also be adopted in this case. Different configurations should be checked while maintaining the $F_s$ adopted in the design. Alternatively, one can opt for a piled embankment solution, as discussed in Chapter 7.

### 6.8   FINAL REMARKS

In stability analyses (in total stress $\phi = 0$) the undrained strength profile $S_u$ may be determined using a number of alternative procedures and for very soft clays piezocone and vane tests are the most cost effective ones. The use of undrained strength equations based on stress history is important to assess the undrained strength profiles obtained. The geosynthetic characteristic should be carefully specified regarding its type (geotextile or geogrid) and the polymer used, as the mobilized geosynthetic strain and the global performance of the reinforced embankment will be influenced by these factors.

Different failure modes should be analyzed, including failures in the embankment body, in the foundation and global embankment-foundation failures. Global failure stability analyses should be performed using different limit equilibrium methods and circular and non-circular failure surfaces. The wedge type failure surface is recommended because it is easy to include reinforcements in the analyses. Wedge analyses may be easily performed by means of spreadsheets with full control of how the reinforcement is used in the calculations, which is not the case with most available softwares for limit equilibrium analyses.

For reinforced embankments, one should evaluate the allowable strain in the reinforcement; the specification of the reinforcement to be used must also be taken into account as well as its stiffness modulus and reduction factors due to mechanical and environmental damage. The reinforcement should be installed as close as possible to the natural terrain in order to provide a greater safety factor $F_s$ in a circular analysis.

In stability calculations of stage-constructed embankments, an in situ evaluation of the undrained strength is recommended before executing the next stage. The stability analyses of staged constructed embankments are optimized when geometry change is incorporated considering previous deformations of the entire embankment-soft soil.

# Chapter 7

# Embankments on pile-like elements

Constructions on very soft soils can result in excessive deformations and stability problems. When this happens, soft soil improvement and embankment stabilization techniques should be evaluated; these can be summarized as shown in Figure 7.1. The techniques listed in the first two columns of the figure were discussed in earlier chapters.

This chapter discusses the techniques listed in the last column of Figure 7.1, i.e., the use of pile-like elements for stabilizing embankments. More specifically, the chapter deals with embankments on piles with geosynthetic reinforcement, embankments over traditional granular columns (most usually stone columns) and embankments on geosynthetic-encased granular columns (most usually sand columns). Other soft ground construction techniques such as jet grouting and deep mixing (using cement) are not in the scope of this book. Embankments on pile-like elements result in shorter construction times compared to other construction methods discussed before the present chapter.

A common feature of the techniques using pile-like elements is the transfer of most of the embankment load to the harder stratum underneath the soft clay layer. Besides this load transfer mechanism, granular columns also promote the increase of the clay strength by radial drainage.

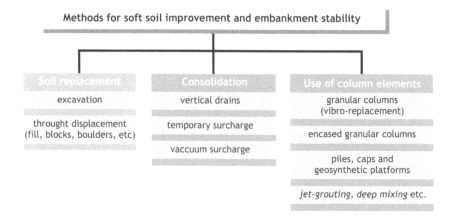

Figure 7.1 Soil improvement and embankment stabilization methods.

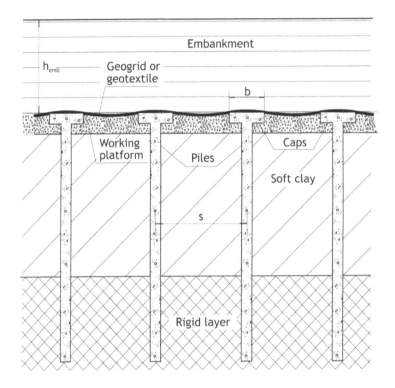

a)  General outline

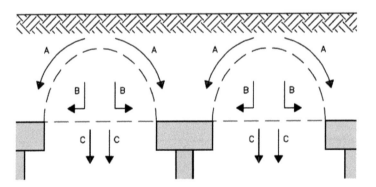

b) Load transfer mechanisms

*Figure 7.2* Embankment on piles reinforced with geosynthetics.

The settlements of embankments on pile-like elements are much smaller than settlements of conventional embankments. Therefore, as far as very soft soils are concerned, the earthmoving volumes are far inferior when compared to the use of conventional embankment.

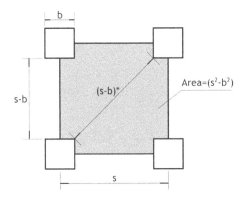

*Figure 7.3* Square caps in square mesh.

## 7.1   PILED EMBANKMENTS WITH GEOSYNTHETIC PLATFORM

The most common type of piled embankment used in very soft clays is shown schematically in Figure 7.2. It consists of a pile driven to the resistant layer underneath the soft clay layer, this pile having usually a pile cap and then a geosynthetic reinforcement layer on its top.

In a piled embankment with geosynthetic platform the embankment load is transmitted by three basic mechanisms (van Eeckelen, 2010): (A) directly to the pile caps through arching effect (Terzaghi, 1943); (B) to the geosynthetic and then to the pile caps, the membrane effect; (C) to the soil in between pile caps under the geosynthetic, which mechanism should not be considered in the case of very soft soils (see 7.1.1). It may be emphasized that, unlike current reinforced embankments, the geotextiles used in piled embankments have to be bidirectional.

The most widely adopted geometry is squares caps in a square mesh, as shown in Figure 7.3 but circular caps and triangular mesh arrangements are also used.

It is observed that the settlement on the surface of the embankment $\Delta h_t$ shown in Figure 7.4 is much smaller than the settlement of the working platform $\Delta h_{if}$. Figure 7.4 also shows that pile caps should ideally have round tops to minimize sharp edges, thus avoiding reinforcement damage.

An important concept related to piled embankments is the "critical height" $h_c$, here defined as the embankment height above which differential settlements at the base of the piled embankment do not produce measurable differential settlement at the embankment surface, i.e., $\Delta h_t = 0$ (see Figure 7.4). Based on field and laboratory as well as analyses of forty four field cases, McGuire et al. (2012) and Filz et al. (2012) found the relationship

$$h_c = 1.15s' + 1.44d \tag{7.1}$$

where $s' = 0.5(s - b)^*$ and $d =$ pile cap diameter $= 1.129b$.

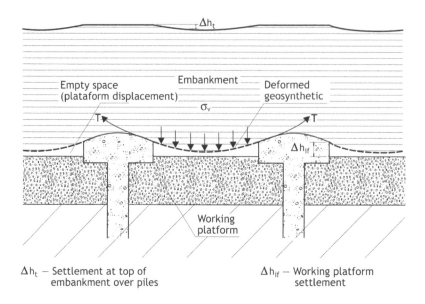

$\Delta h_t$ – Settlement at top of
embankment over piles

$\Delta h_{if}$ – Working platform
settlement

*Figure 7.4* Piled embankment: settlements, vertical stress and tensile force on geosynthetic.

It should be pointed out that the critical height has also been referred (e.g., Horgan and Sarsby, 2002) as the height above which all additional loads due to fill and surcharge are distributed completely to the pile caps.

The detailed design at the embankment slope requires a separate study (BS 8006 – 2010) regarding proper anchoring of the reinforcement laterally. Numerical analysis by finite elements may be adopted in this case (Almeida, Almeida and Marques, 2008; Gebreselassie, Lüking and Kempfert, 2010; Jennings and Naughton, 2010), but this topic will not be discussed here.

### 7.1.1 The working platform settlement and overall embankment behavior

The construction of the working platform (see Chapter 1) is the first step to allow access of pile driving equipment on soft clay deposits without any fill layer on the ground surface. Then piles are driven and caps installed. The caps may be placed above or inside the working platform, as shown in Figure 7.5A,B, and then the geosynthetic is installed above the caps. It is observed that in any of the cases, the working platform will suffer consolidation due to settlements (primary and secondary compression), as shown in Figure 7.5C,D (Almeida et al., 2008a). For this reason, the reaction of the soil below the geogrid, which is considered in certain calculation methods (e.g., Kempfert et al., 2004), will not be taken into account here.

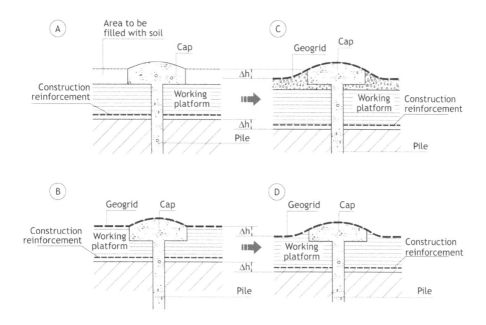

Figure 7.5 Details of the placement of caps on piled embankments: (A) and (C) cap above the working platform; (B) and (D) cap embedded in the working platform.

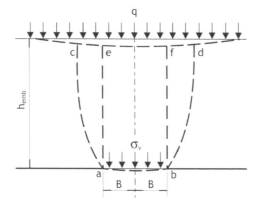

Figure 7.6 Model for the study of the arching effect on the soils (Terzaghi, 1943).

## 7.1.2 Arching effect on soils

An important phenomenon for the study of piled embankments, with or without a geogrid platform, is the arching effect on the soil (Terzaghi, 1943) and outlined in Figure 7.6. In these studies, Terzaghi considered the condition of two-dimensional (plane strain condition), but the most common condition of a piled embankment is three-dimensional.

By analyzing the equilibrium in the vertical direction of a soil element in line with span 2B, where 2B is the distance between caps $(s - b)$, Terzaghi obtained the value of vertical stress acting on the base of the embankment $\sigma_v$:

$$\sigma_v = \frac{(s-b)\left(\gamma_{emb} - \frac{c_{emb}}{(s-b)}\right)}{Ktg\,\phi_{emb}}\left(1 - e^{-Ktg\,\phi_{emb}\frac{h_{emb}}{(s-b)}}\right) + q.e^{-Ktg\,\phi_{emb}\frac{h_{emb}}{(s-b)}} \qquad (7.2)$$

where:
$c_{emb}$ – cohesion of the fill material $(kN/m^2)$;
$\phi_{emb}$ – internal friction angle of the fill material $(°)$;
$K$ – earth pressure coefficient of fill material; $K = 1$ is recommended (numerical studies by Potts and Zdravkovic, 2010);
$s - b$ – distance between caps (m);
$\gamma_{emb}$ – specific weight of the fill material $(kN/m^3)$;
$q$ – uniform surcharge on the surface per unit area $(kN/m^2)$;
$h_{emb}$ – height of the embankment (m).

### 7.1.3 Defining the geometry of piled embankments

The first step in the design of the piled embankment is the definition of the geometry of the problem (spacing s, width of the pile cap b; height of embankment $h_{emb}$).
The British Code BS8006 (2010) recommends:

$$h \geq 0.7(s - b) \qquad (7.3)$$

The German code EBGEO (2010) recommendations are based on the span $(s - b)^*$ defined by the diagonal distance between caps $(45°)$, which for static loads are

$$h \geq 0.8(s - b)^* \qquad (7.4)$$

and

$$(s - b)^* \leq 2.5\,m \qquad (7.5)$$

And in the case of impact loads

$$h \geq 2.0(s - b)^* \qquad (7.6)$$

$$(s - b)^* \leq 3.0\,m \qquad (7.7)$$

EBGEO (2011) also recommends

$$b/s > 0.15 \qquad (7.8)$$

The Dutch code (van Eekelen et al., 2010) recommends:

$$h \geq 0.66(s - b) \qquad (7.9)$$

With regard to the fill material, the Dutch code recommends materials with $\phi_{emb} \geq 35°$ for the height range of geosynthetic above the embankment corresponding to $h_{emb} \leq 0.66(s - b)^*$ and $\phi_{emb} \geq 30°$ for $h_{emb} \geq 0.66\ (s - b)^*$.

The Dutch code also recommends that the geosynthetic should not be directly supported on the pile cap, but on a granular layer of soil above it. If there is only one layer of geosynthetic, this layer should have a distance z from the pile cap of up to 0.15 m. In the case of a second layer of geosynthetic placed above, the distance between this layer and the first layer below should be less than 0.20 m. The arching effect changes when using more than one layer of geosynthetic (Gebreselassie, Lüking and Kempfert, 2010).

Blanc et al. (2013) studied piled embankments by means of centrifuge tests in which a mobile tray device simulated the settlement of the soft soil located between the inclusions. A parametric study of the load-transfer mechanisms within the embankment was conducted for three different fill thicknesses h and two pile span values (s − b). It was found that load transfer mechanisms are better for thicker fills and for lower pile span values (s − b), a finding valid with or without geosynthetic reinforcement. The improvement made by a geosynthetic reinforcement was clearly shown through both load transfer and differential settlement reduction.

By defining the pile spacing and pile cap dimensions for a given embankment height, the next calculations to be made are the vertical stress acting on the geosynthetic and the geosynthetic tension force, and these two calculations are presented below.

### 7.1.4  Calculation of vertical stresses acting on the geosynthetic

Several methods are available for the design of piled embankments. The methods proposed by Russell and Pierpoint (1997), Hewlett and Randolph (1988), Kempfert et al. (2004), Filz and Smith (2006) and BS 8006 (2010) are among the most widely used. The methods described below were chosen based on their simplicity and consistency.

#### Method of Russell and Pierpoint (1997)

Russell and Pierpoint (1997) adapted the Terzaghi method and considered $K = 1$, in order to take into account the three-dimensional nature of the pile arrangement. The equation is adopted by these authors is:

$$\frac{\sigma_v}{(\gamma_{emb}h_{emb} + q)} = \frac{s^2 - b^2}{4h_{emb} \cdot b \cdot K \cdot tg\ \phi_{emb}} \left\{ 1 - e^{\frac{4h_{emb} \cdot b \cdot K \cdot tg\ \phi_{emb}}{s^2 - b^2}} \right\} \tag{7.10}$$

This method does not consider the reaction of the soft soil underlying the geosynthetic. This reaction, however, is not relevant in the case of very soft clays, as mentioned before.

#### Method by Kempfert et al. (2004) used in EBGEO (2011)

Kempfert et al. (2004) presented a method for calculating the vertical stresses in the reinforcement using an analytical model of a dome based on the theory of elasticity, and charts (for $\phi = 30°$), which allows the calculation of the vertical stresses, as shown in Figure 7.7. This method has been adopted by EBGEO (2010), which presents charts

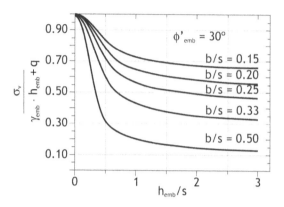

*Figure 7.7* Calculation of vertical stresses on the reinforcement (Kempfert et al., 2004).

for other values of the fill material friction angle, besides presenting also equations to allow calculation of the vertical stresses in the reinforcement for any given geometry and parameters.

## 7.1.5   Calculation of tensile force acting on the reinforcement

The most used methods (McGuire and Filz, 2008) used to calculate the geosynthetic tensile force due to the vertical stress acting (Figure 7.3) on the reinforcement are: (a) the parabolic method; (b) the tensioned membrane method; (c) the method by Kempfert et al. (2004), which are summarized below.

Some authors calculate the value of T from an assumed or prescribed reinforcement strain ($\varepsilon$) which not necessarily produces consistent values (McGuire and Filz, 2008). Therefore, the equations and methods described below are computed adopting the reinforcement modulus J, not a prescribed value of $\varepsilon$. In the parabolic method the stress at the reinforcement T is calculated, assuming that the deformation of the reinforcement in the pile span (s − b) has a parabolic shape (Figure 7.3). The value of T is then given by equation (7.11) according to the value of the reinforcement modulus J to be used (McGuire and Filz, 2008):

$$96T^3 - 6\hat{K}_g^2 T - \hat{K}_g^2 J = 0 \tag{7.11}$$

where

$$\hat{K}_g = \frac{\sigma_v(s^2 - b^2)}{b} \tag{7.11a}$$

and $(s^2 - b^2)$ is the area shown in Figure 7.3.

The tensioned membrane method (Collin, 2004) is an adaptation of the work by Giroud (1990) for calculating tension in geosynthetics spanning voids. This procedure assumes a circular void exists beneath the reinforcement with a diameter equal to the diagonal clear spacing, $\sqrt{2} \cdot (s - a)$, for a square array of columns. Tensioned membrane theory assumes the deflected shape of the reinforcement is a circular arc

Table 7.1 General characteristics of some piled embankments in Brazil (Huesker).

| Location (Brazilian State) | Embankment Height h (m) | Piles spacing s (m) | Pile cap width a (m) | Fortrac Geogrid | $\dfrac{h}{(s-a)}$ | Area ratio $-a_c$ (*) |
|---|---|---|---|---|---|---|
| Paraná | 3 | 2.7 | 1 | 150/150 | 1.76 | 10.77% |
| Minas Gerais | 11 | 1.7 | 0.8 | 110/110 & 200/200 | 12.70 | 17.39% |
| São Paulo | 5.5 | 2.4 | 1.1 | 100/100 | 4.23 | 16.50% |
| Brasília – DF | 6 | 2 | 0.5 | 100/100 | 4.00 | 4.91% |
| São Paulo | 1.9 | 2 | 0.8 | 400/100 & 300/100 Fortrac 200/100 | 1.58 | 12.57% |
| São Paulo | 4.2 | 2 | 1.1 | 400/100; 300/100; Fortrac 200/100 | 4.67 | 23.76% |
| Rio Grande do Sul | 2.4 | 2.3 | 0.9 | 200/200 | 1.71 | 12.03% |
| Rio Grande do Sul | 3.6 | 2.3 | 0.9 | 200/200 | 2.57 | 12.03% |
| Rio/SENAC | 1.2 | 2.5 | 0.8 | 200/200 | 0.71 | 8.04% |
| Rio/SESC | 1.4 | 2.8 | 1 | 200/200 | 0.78 | 10.02% |

(*) $a_c = b^2/s^2$ for a squared pile and square caps arrangement (see Figure 7.3 for b and s).

(Figure 7.3). Knowing the value of the modulus J, the value of T is defined by the equation:

$$\frac{2\sqrt{2}\cdot T\cdot J}{\sigma_v(s-b)}\cdot \text{sen}^{-1}\left[\frac{\sigma_v(s-b)}{2\sqrt{2}\cdot T}\right]-T-J=0 \tag{7.12}$$

Kempfert et al. (2004) and EBGEO (2010) present charts that consider the possibility of favorable contribution of the soil reaction below the reinforcement, which is not recommended in the case of very soft clays, as discussed previously.

McGuire and Filz (2008) presented parametric studies comparing the parabolic and the tensioned membrane methods and found that the parabolic method results in greater tensile force values than the tensioned membrane and Kempfert methods.

Other design methods have been presented in the literature. For instance, the method by Abusharar et al. (2009), derived from Low et al. (1994)'s method, considers the soft soil as linear elastic, but equations can be manipulated so that the soft soil reaction does not exist. The method allows calculation of the vertical displacement under the geosynthetic as well as the geosynthetic tension force T.

### 7.1.6 Case histories of piled embankments

Recent case histories of piled embankments have been presented in Briançon, Delmas and Villard (2010); van Eekelen, Bezuijen and Alexiew (2010), van der Stoel et al. (2010) and ASIRI (2012). Some applications in Brazil are summarized in Table 7.1.

With respect to applications in very soft soils, Almeida et al. (2008a) describe the behavior of two low piled embankments constructed in Barra da Tijuca (RJ, Brazil). Instruments were installed a test area to conduct a study on the behavior of the embankment-reinforcement-cap pile system (Almeida et al., 2007). The test area

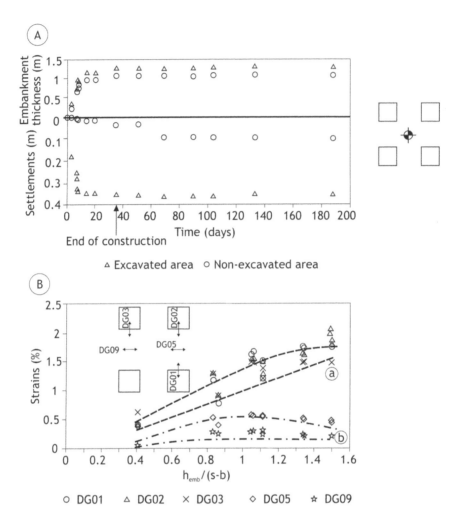

*Figure 7.8* (A) Measured settlements in the center of excavated and unexcavated areas (3D arrangement – section 1), (B) Deformation measurements (in the excavated area) on geogrid in points: (a) on the face of the capital and (b) half distance between caps (Spotti, 2006; Almeida et al., 2007).

included a conventional three-dimensional layout (square pile caps) as well as a two-dimensional layout (linear pile caps). An excavation was performed below the geogrid, in order to accelerate the load transfer to the reinforcement. The settlements measured between the caps (just above the geosynthetic) in the three-dimensional layout were in the order of 0.1 m to 0.4 m (Figure 7.8A). The strains in the reinforcement were in the range of 0.25% to 2.0%, depending on the point of measurement (Figure 7.8B). McGuire, Filz and Almeida (2009) made predictions of settlements of embankment 1 by means of the Filz and Smith (2006) method and found good agreement with field monitoring data. Sandroni and Deotti (2008) also reported data of another test embankment nearby.

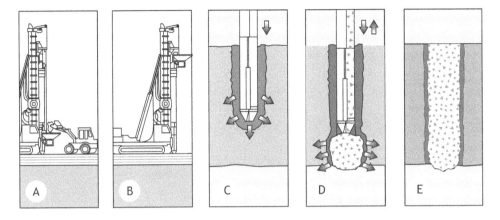

*Figure 7.9* Execution sequence of a gravel column in saturated soft soil (McCabe, McNeill and Black, 2007).

## 7.2   EMBANKMENTS ON TRADITIONAL GRANULAR COLUMNS

One of the most used methods for soft soil improvement is the use of granular columns with sand or gravel. The granular columns can be installed with or without lateral displacement of the clay around it. The columns installed with clay displacement (closed-ended tube) are commonly used, especially if the columns are installed by means of vibro-replacement. Thus, a network of granular columns is produced which transfers most of the embankment load to the stiffer underlying layer. Granular columns decrease embankment settlements and improve overall stability. The granular columns also allow the dissipation of pore pressures by radial drainage, thus increasing the clay strength and accelerating settlements.

Floating granular columns, i.e., not resting on a hard stratum below the clay layer, may also be used (Weber et al., 2009, Gäb et al., 2009) and in this case the clay-column side friction is an important mechanism for load transfer rather than point bearing.

### 7.2.1   Traditional granular columns using the vibro-replacement method

Traditional columns have been installed since the mid-twentieth century, often through the vibro-replacement technique (e.g. Baumann and Bauer, 1974; Raju and Sonderman, 2005; Raju, Wegner and Godenzie, 1998). The granular material used in the columns is usually gravel (crushed rock), but sand columns are also used, particularly in Japan (Kitazume, 2005).

Figure 7.9 shows the execution phases of a granular column by the vibro-displacement process. Initially, a bucket is filled with granular material (Figure 7.9A), which is then lifted and fills the tube with the granular material (Figure 7.9B). The tube penetrates the soil by jetting thus increasing the tube diameter (Figure 7.9C) and once the desired depth is reached, the granular material is introduced into the borehole (Figure 7.9D). Through upward and downward movements of the tube, the granular

*Table 7.2* Literature recommendations (adapted from FHWA, 1983).

| Conditioning Factors | Recommendations |
|---|---|
| % of soft clay going through the 200 sieve | less than 15% to 30% |
| $S_u$ of soft clay | Between 15 kPa up to 50 kPa (*) |
| Diameter of columns | 0.6 m to 1.0 m |
| Spacing between columns | 1.5 m to 3.0 m |
| Length of columns | Between 6 m up to 10 m (*) |
| Grain diameter of the column material | 20 mm to 75 mm |
| Friction angle of the granular soil | 36° to 45° |
| Stone column Young's modulus | 30–100 MPa (lower range for design) |

(*) FHWA (1983) reports cases with $S_u$ values as low as 7.5 kPa and column lengths up to 20 m.

material is compacted, while more material is introduced into the preformed hole. In addition, the jetting is carried out to ensure the formation of a column with clean granular material (Figure 7.9C, D). This operation is performed up to the ground surface, until the column is completely formed (Figure 7.9E).

Table 7.2 lists parameter for the proper performance of vibro-replacement technique based on the experience of more than fifty years (Greenwood, 1970; Thornburn, 1975; FHWA, 1983) with the traditional stone columns technique.

### 7.2.2   Design and analysis principles

Similarly to the case of vertical drains, granular columns with diameter d can be installed with spacing l in square or triangular meshes (see Figure 5.4).

*Unit cell concept*

Most methods of granular columns design are based on the unit cell concept (Figure 7.10), with diameter equivalent $d_e = 1.13l$ or $d_e = 1.05l$, respectively, in the case of square or triangular mesh, and the column area, $A_c = \pi d^2/4$; the total cell area, $A = \pi d_e^2/4$, and the area of the soft soil around the column, $A_s = A - A_c$. Thus, the area replacement ratio is defined by:

$$a_c = \frac{A_c}{A} = c \cdot \left(\frac{d_e}{l}\right)^2 \tag{7.13}$$

where c is equal to $\pi/4$ and $\pi/(2\sqrt{3})$, respectively for rectangular and triangular mesh. The soft soil area ratio is then defined by

$$a_s = \frac{A_s}{A} = 1 - a_c \tag{7.14}$$

*Column and soil vertical stress distribution*

Studies indicate that when the soil-column system is loaded, a stress concentration occurs in the columns, due to the greater column stiffness compared to the surrounding soft soil, thus arching develops. The stress concentration factor n is expressed by the

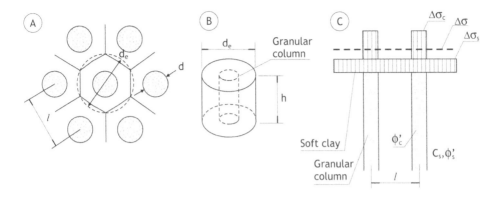

*Figure 7.10* Unit cell scheme: (A) top view, (B) unit cell, (C) stress distribution.

ratio between the increments of vertical stresses acting in the column $\Delta\sigma_c$ and in the soft clay around it $\Delta\sigma_s$ (both defined in Figure 7.10c):

$$n = \frac{\Delta\sigma_c}{\Delta\sigma_s} \qquad (7.15)$$

Numerical studies correlated the stress concentration factor n to the ratio between the elasticity modulus of the column $E_c$ and elasticity modulus of the clayey soil $E_s$ (FHWA, 1983). The results can be expressed by the equation (Han, 2010):

$$n = 1 + 0.217\left(\frac{E_c}{E_s} - 1\right) \qquad (7.16)$$

Han (2010) recommended $E_c/E_s$ values lower than 20, since higher values are not mobilized in situ, although they may be measured in the laboratory. For $E_c/E_s = 20$, $n = 5$ is obtained, which should be the maximum value of n.

A large number of experimental and numerical studies (Barksdale and Bachus, 1983; Mitchel and Huber, 1985; Kitazume, 2005; Castro, 2008; Murugesan and Rajagopal, 2010; Choobbasti et al., 2011; Six et al., 2012) have addressed the stress concentration factor n. Based on these studies the recommended n values for gravel and sand columns are between 2 and 5. These values of n refer to the top of the column and for long term conditions. Numerical analyses (Lima, 2012) show that for any given time n varies with time. Besides n also varies with depth, as a mechanism similar to arching also occurs with depth.

The applied embankment vertical stress $\Delta\sigma$, equal to the specific weight of the embankment $\gamma_{emb}$ times the height of the embankment $h_{emb}$, is shared between column (vertical stress acting on top $\Delta\sigma_c$) and soft soil (vertical stress acting on top $\Delta\sigma_s$). The equilibrium of the vertical forces within the unit cell yields

$$\Delta\sigma A = \Delta\sigma_c \cdot A_c + \Delta\sigma_s \cdot A_s \qquad (7.17)$$

By dividing both sides by A:

$$\Delta\sigma = \Delta\sigma_c \cdot a_c + \Delta\sigma_s \cdot (1 - a_c) \qquad (7.18)$$

Substituting Equation (7.15) in (7.18) yields

$$\Delta\sigma_s = \frac{\Delta\sigma}{[1 + (n-1)a_c]} = \mu_s\Delta\sigma \qquad (7.19)$$

$$\Delta\sigma_c = \frac{n \cdot \Delta\sigma}{[1 + (n-1)a_c]} = \mu_c\Delta\sigma \qquad (7.20)$$

### 7.2.3  Settlement reduction factor (soil improvement factor)

The classic methods used to calculate settlements in stone columns use generally the concept of settlement reduction factor β, defined by the ratio between the soft soil settlement Δh without treatment and the soft soil settlement $\Delta h_s$ with treatment, or

$$\beta = \frac{\Delta h}{\Delta h_s} \qquad (7.21)$$

Settlements on the unimproved (untreated) soft soil can be expressed by the coefficient volumetric compressibility of the soil $m_v$:

$$\Delta h = h \cdot m_v \cdot \Delta\sigma \qquad (7.22)$$

Where $m_v$ is equal to the inverse of the oedometer modulus, $E_{oed}$ or $m_v = 1/E_{oed}$, h is the clay thickness and Δσ is the applied embankment stress (see Figure 7.10). Assuming that the settlements of the soil-column system are due solely to the soft soil settlement, one can express it by:

$$\Delta h_s = h \cdot m_v \cdot \Delta\sigma_s \qquad (7.23)$$

Substituting (7.22) and (7.23) in (7.21)

$$\beta = \frac{\Delta\sigma}{\Delta\sigma_s} \qquad (7.24)$$

And using equation (7.19)

$$\beta = 1 + (n-1)a_c \qquad (7.25)$$

The method proposed by Priebe (1978, 1995) is the most used method to estimate the magnitude of the settlements, as discussed below.

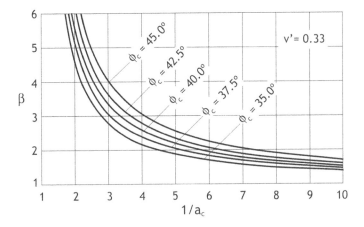

*Figure 7.11* Settlement reduction factor versus area replacement ratio.

## 7.2.4 Settlement computations

### Magnitude of settlements

For the evaluation of settlements of an embankment on granular columns, Priebe (1995) proposed a calculation method where the following hypotheses are considered:

- The column is based on a rigid layer;
- The column material is uncompressible;
- The bulk density of column and soil is neglected;
- The soil is assumed to be displaced already during the column installation to such an extent that its initial resistance corresponds to the liquid state, i.e. the coefficient of earth pressure amounts to $K = 1$.
- The column material shears from the beginning whilst the surrounding soil reacts elastically.
- the column rests on a stiff layer;
- soil and column settlements are the same;
- specific weights of the soil and column are disregarded;
- the cross-sectional area of the unit cell remains constant.

Based on these hypotheses, equations for the value of β have been developed (Priebe, 1995), which for the case of Poisson's coefficient $v' = 0.33$, is expressed by:

$$\beta = 1 + a_c \left[ \frac{(5 - a_c)}{[4K_{a,c}(1 - a_c)]} - 1 \right] \tag{7.26}$$

Figure 7.11 express Eq. (7.26) graphically.

Priebe (1995) also presents charts and calculation procedures that incorporate the compressibility of the columns and bulk weights of the column and soil. Based on monitoring data of eighteen projects using stone columns McCabe (2009) found

*Chart 7.1* Theoretical solutions for calculating the settlement rate.

| Source | Disturbance and well resistence | Permeability Coefficient Region remolded | Load type |
|---|---|---|---|
| Balaam and Booker (1981) | No | – | immediate |
| Han and Ye (2001) | No | – | immediate |
| Han and Ye (2002) | Yes | Unique value | immediate |
| Castro (2008); Castro and Sagaseta (2009) | No | – | immediate |
| Wang (2009) | Yes | Unique value | variable |
| Xie et al. (2009a) | Only Disturbance | 3 patterns | immediate |
| Xie et al. (2009b) | Yes | Variable zone remolded, Standard 2 of Xie et al. (2009a) | variable |

good agreement between measured and computed values of the improvement factor β given by Priebe's method. The adopted value of friction angle of the column material in all projects was $\phi = 40°$, a typical value used in design. Mitchell & Huber (1985) and Springman et al. (2012) also report good agreement between measurements and Priebe's method.

Other calculation methods have been more recently proposed (Bouassida et al., 2003; Pulko and Majes, 2005; Castro and Sagaseta, 2009) but were not yet sufficiently assessed.

Mestad et al. (2006) report results of a prediction symposium in which seventeen groups predicted settlements of a field test. In general, the predictions presented a high discrepancy in results, even when carried out by the same calculation method with differences of almost 100% on settlements values at the embankment centerpoint. Mestat et al. (2006) concluded that either complex 3D finite element models or simple one-dimensional approach, with a "good" choices of parameters, may result in adequate settlement predictions.

Numerical methods, in particular finite elements, have been extensively used at present (e.g., Kirsch and Sondermann, 2001; Gäb et al., 2009; Weber et al., 2009; Castro and Karstunen, 2010) in order to allow a number of possibilities not available in analytical methods. However, the adequate use of numerical methods is not trivial and require due consideration of the column installation effects (Egan et al., 2008) and also 2D versus 3D correspondence as well as consolidation analyses.

### Settlement rate

Several authors have presented theoretical solutions for calculating the variation of settlement with time and a compilation of some important studies is presented in Chart 7.1.

The clay penetration in the column during its installation, using the vibro-replacement method, may results in substantial reduction in the permeability coefficient of the column material (Han, 2010). Even so, for usual adopted column replacement ratio values $a_c$, the settlements of embankments on granular columns usually stabilize relatively fast, most often in a matter of a few months (Tan, Tjahyono and Oo, 2008).

## 7.2.5   Stability analysis

In general, the stability analysis of embankments on granular columns is performed on the basis of the composite ground parameters, i.e., the strength parameters $c_m$ and $\phi_m$ and the specific weight $\gamma_m$ of the soil-column system. These are calculated from the strength parameters of the soft clay ($c_s = S_u$ and $\phi_s = 0$; relevant for undrained total stress analysis) and the granular column ($\phi_c$), and the stress concentration parameter m. Priebe (1995)'s method is the most currently used for this purpose and the values of $c_m$ and $\phi_m$ are determined by:

$$\text{tg } \phi_m = m \text{ tg } \phi_c + (1 - m)\text{tg } \phi_s \qquad (7.27)$$

$$c_m = (1 - m)c_s \qquad (7.28)$$

The parameter m is calculated from the relative stress distribution between column-soil, defined from the parameters n (Eq. 7.15) and $a_c$ (Eq. 7.13), according to:

$$m = a_c \frac{\Delta\sigma_c}{\Delta\sigma} = a_c\mu_c \qquad (7.29)$$

Or substituting (7.20) in (7.29):

$$m = \frac{a_c n}{[1 + (n - 1)a_c]} \qquad (7.30)$$

Priebe (1995) developed graphics based on $a_c$ and $\phi'_c$ for the rapid determination of m. Di Maggio (1978) recommended a lower limit of $m = a_c$. The average specific weight of the improved soil can be expressed as:

$$\gamma_m = \gamma_c \, a_c + \gamma_s(1 - a_c) \qquad (7.31)$$

Alternatively, the stability analysis may be performed directly without using the concept of composite ground, i.e., granular columns and soft soil are both materialized in the embankment foundation (Kitazume, 2005). In this more classical type of analyses due correspondence has to be done regarding the 3D versus plane strain analyses, as the circular column changes to a trench (Tan et al., 2008).

## 7.2.6   General behavior of embankments on granular columns

### Construction of embankments

Studies of physical modeling in a centrifuge by Almeida (1984) and Almeida, Davies and Parry (1985) compared the performances of two embankments with granular

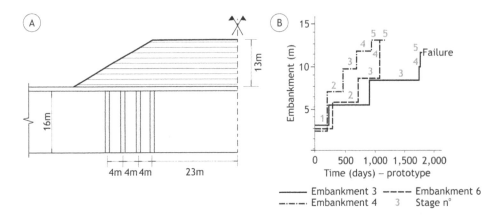

*Figure 7.12* Prototype centrifuge model: (A) geometry of the embankment 6, (B) loading history of embankments 3, 4 and 6 (Almeida, 1984).

columns (embankments 4 and 6) installed with an open tube (soft clay previous removed), with an embankment without granular columns (embankment 3).

In such cases the columns were installed only under the slope of the embankment (with the main purpose of increasing stability), as illustrated in 7.12A (prototype). This alternative, which is more economical than the reinforcement of all the clay under the embankment, has been used a few times (Rathgeb and Kutzner, 1975; Yee and Raju, 2007).

Two $a_c$ area substitution ratios (Eq. 7.13) of the clay were used in this case: 4.9% (embankment 4, $s = 3.0$ m, $d = 1.0$ m) and 8.7% (embankment 6, $s = 4.0$ m, $d = 1.0$ m). These values are lower than those commonly used in very soft clays (generally between 10% and 20%), but were intended to evaluate the worst case scenario for the effectiveness of granular columns for ground improvement in soft soils.

The three embankments were built in five stages, with loading histories indicated in Figure 7.12b (scale of prototype). Embankment 3 had a lower final height (which was the failure height) than the other two embankments with columns (without failure) and also required a much longer construction period (unrealistic in terms of engineering practice).

### Horizontal displacements

The results of horizontal displacements $\delta_h$ for embankments 3 and 6 are shown in Figure 7.13, for inclinometer I1 for the positions indicated in the figure. Comparing the displacements in the foundations of embankments 3 and 6, we conclude that:

- Embankment 6, with granular columns, presented about half the displacements to those of embankment 3, without granular columns, and reached the greatest height in the shortest construction period;
- Embankment 3 presented a more drastic variation of displacements according to depth than embankment 6, particularly after layer 3. The displacements of

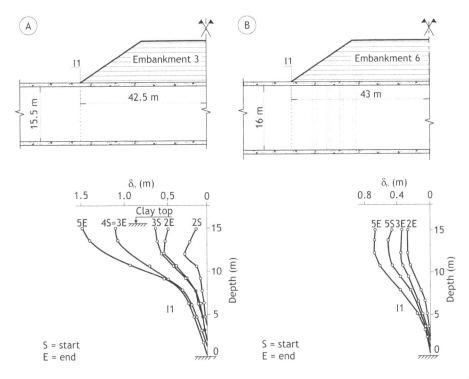

*Figure 7.13* Geometry and inclinometric profile I1 (prototype centrifuge model) (A) embankment 3, (B) embankment 6 (Almeida, 1984).

the embankment by the end of stage 5 refer to the failure moment, i.e. they are excessive and incompatible with the behavior of the embankment in service conditions.

## Vertical displacements

The vertical displacements $\Delta h$ at the base of embankments 3, 4 and 6 are compared in Figure 7.14, at the end of loading stage 5 (see Figure 7.12), which for embankment 3 it is the moment immediately before failure. It is observed that the settlements in embankments 4 and 6 are quite similar, i.e. the short loading history used in embankment 4 compensated the smallest granular pile spacing. In embankment 3, the settlements and the heave in front of the embankment are clearly superior to the others, confirming the results of horizontal displacements.

## Measured pore pressures

The piezometers of, embankment 6, equipped with granular columns under the slope, presented faster dissipation of excess pore pressures than piezometers of embankment 3 (Figure 7.15). Note, for example, the piezometer P7 located within the region of the

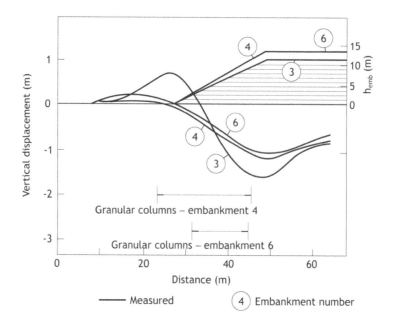

*Figure 7.14* Vertical displacements at the base of embankments 3, 4 and 6 at the end of stage 5 (Almeida, 1984).

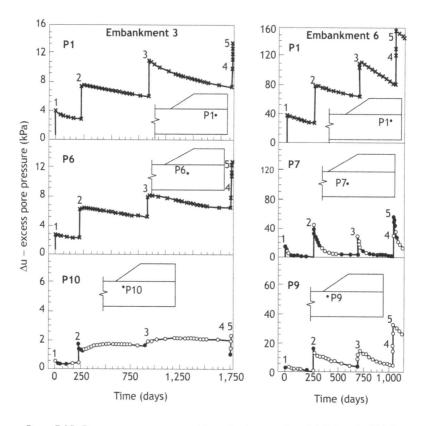

*Figure 7.15* Pore pressures measured in embankments 3 and 6 (Almeida, 1984).

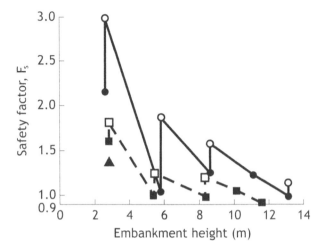

**Symbols:**

| Embankment 3 | ■ Beginning of stage<br>□ End of stage | Effective stress analysis |
|---|---|---|
| | ▲ S$_u$ vane | Total stress analysis |
| Embankment 6 | ● Beginning of stage<br>○ End of stage | Effective stress analysis |

*Figure 7.16*  F$_s$ variation during construction of embankments 3 and 6 (Almeida, 1984).

sand piles of embankment 6, which had a faster dissipation rate than P6 in embankment 3. Also the piezometer P1 relatively far from the piles in embankment 6, also had a faster dissipation rate than P1 in embankment 3.

Stability analysis: Staged embankments with and without granular columns.

The stability analyses of embankments 3 and 6 were performed in terms of effective stress, using the measured the pore pressures during the staged construction (some measured values are shown in Figure 7.15).

A comparison is presented in Figure 7.16, between the safety factors F$_s$ of embankments 3 and 6. As expected, much greater values of F$_s$ are observed – and also F$_s$ increases in each stage – for the foundation treated with granular columns. Embankment 6 was approximately 13 m in height without failure, the final F$_s$ value around unit. Embankment 3 failed at 11.6 m high, with F$_s$ equal to 0.91.

The better results of stability analyses, and maximum height reached by embankment 6 with granular columns, compared with embankment 3 without columns, clearly indicates the beneficial effect of the granular columns as a soft clay ground improvement technique.

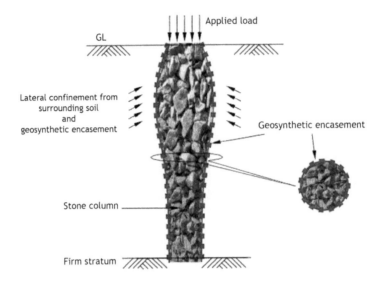

*Figure 7.17* Schematic of geosynthetic encased stone column (Murugesan and Rajagopal, 2006).

## 7.3  ENCASED GRANULAR COLUMNS

### 7.3.1  General description

When the stone columns are installed in extremely soft soils, they may not provide significant load capacity owing to low lateral confinement. McKenna et al. (1975) reported cases where the stone column was not restrained by the surrounding soft clay, which led to excessive bulging, and also the soft clay squeezed into the voids of the aggregate. The squeezing of clay into the stone aggregate ultimately reduces the bearing capacity of stone column. Also the lower undrained cohesion value demand more stone column material.

The use of traditional stone columns is limited to certain values of undrained cohesion of the soft soil (see Table 7.2), thus confining the granular column with a high-stiffness geosynthetic encasement (Raithel et al., 2000; Alexiew et al., 2005; di Prisco et al., 2006; Murugesan and Rajagopal, 2006) could overcome this difficulty.

Van Impe and Silence (1986) were probably the first to recognize that columns could be encased by geotextile. They produced an analytical design technique that was used to assess the required geotextile tensile strength. Details on this technique were provided by Kempfert et al. (1997). Raithel and Kempfert (2000) produced an analytical design technique for assessing column settlement based on geotextile stiffness. An update, including use on recent projects in Europe, was provided by Raithel et al. (2005) and Alexiew et al. (2005) and in South America by Mello et al. (2008). The general idea of Geosynthetic Encased Columns (GEC) is shown in Figure 7.19. Also, the principle of this technique, developed in Germany in the mid-1990s, is shown in Figure 7.18.

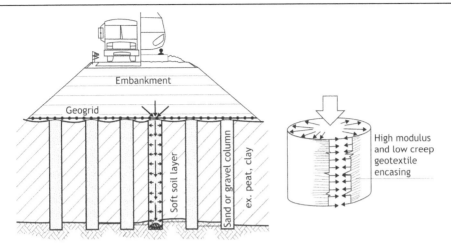

*Figure 7.18* Outline of an embankment built on soft soil over columns encased with geosynthetic (Raithel and Kempfert, 2000).

In general, the encasing used for the columns consists of a woven geotextile with high modulus and low creep coefficient, which maintains the favorable drainage characteristics of the granular column. The granular material used can be sand or gravel; the latter, however, provides higher overall column stiffness.

The geosynthetic encasing ensures the nominal column diameter, minimizes material losses, increases overall column stiffness and avoids granular column contamination preserving its permeability.

## 7.3.2   Execution methods

Encased columns can be executed with or without lateral displacement of the clay thus two different methods are generally available with regards to the GEC construction technology.

The first technique is the displacement method where a closed-tip steel pipe is driven down into the soft soil followed by the insertion of the circular woven geotextile and sand or gravel backfill (Figure 7.19). Finally, the pipe is pulled upwards under optimized vibration designed to compact the column and then the column is completed.

The displacement method is commonly used for very soft soils (e.g. $<15\,\mathrm{kN/m^2}$).

Encased columns have usually a diameter of approximately 0.80 m, the diameters of geotextile casing and tube (inside) being the same (Alexiew, Horgan and Brokemper, 2003). The column spacing is typically between 1.5 m and 2.5 m and the stiffness modulus (J) of the geotextile generally varies between 1500 kN/m and 6500 kN/m.

The second construction technique used for encased columns is the replacement method with excavation of the soft soil inside the pipe. With the replacement method, an open steel tube is driven deep into the bearing layer and the soil within the tube is removed out by auger boring as shown in Figure 7.20. The replacement method is preferred for soils with relatively higher penetration resistance or when vibration effects on nearby buildings and road installation have to be minimized.

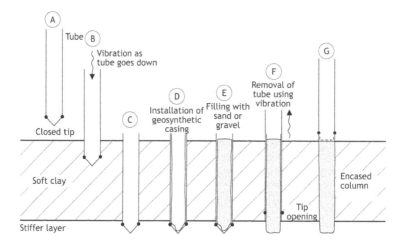

Figure 7.19  Displacement method for GEC installation (Alexiew, 2005).

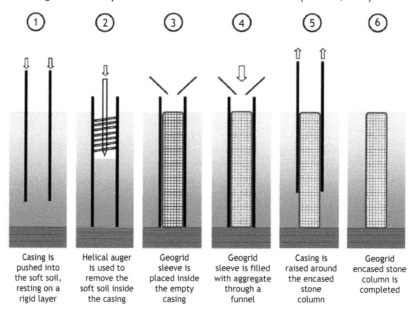

Figure 7.20  Replacement Method (Gniel and Bouazza, 2010).

### 7.3.3  Calculation methods

The most widely used calculation methods for encased columns design are the ones proposed by Raithel (1999) and Raithel and Kempfert (2000). The main hypotheses of these methods for a unit cell with radius $r_e$ are (refer to Figure 7.19):

- column installed in underlying non-deformable layer;
- equal settlements for the column and soil around it;

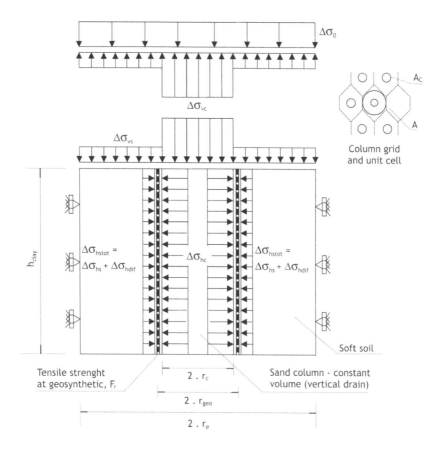

*Figure 7.21* Calculation model of the granular encased column.

- active earth pressures $K_{ac}$ on the column;
- for the excavation method, $K_{os} = 1 - \sin \phi'$ is used for the soil surrounding the granular column; for the displacement method, $K_{os}^*$ (increased $K_o$) is used;
- the geotextile reinforcement has a linear elastic behavior;
- calculation for the drained condition (soft clay with parameters $c_s'$ and $\phi_s'$), because this is the condition for larger settlements.

The geosynthetic responsible for the encasing (cylinder radius $r_{geo}$) has a linear-elastic behavior and stiffness modulus J, and the increase in tensile force of the geosynthetic is given by:

$$\Delta F_R = J \cdot \frac{\Delta r_{geo}}{r_{geo}} \tag{7.32}$$

The compatibility of horizontal deformations relates the value of the column radius variation ($\Delta r_c$) with the variation of the geosynthetic radius ($\Delta r_{geo}$) according to Eq. (7.33), where $r_c$ is the initial column radius:

$$\Delta r_c = \Delta r_{geo} + (r_{geo} - r_c) \tag{7.33}$$

The variation of the column radius ($\Delta r_c$) is calculated by means of the approach proposed by Ghionna and Jamiolkowsky (1981) according to the difference of horizontal stresses $\Delta\sigma_{hdif} = \Delta\sigma_{hc} - (\Delta\sigma_{hs} + \Delta\sigma_{hgeo})$, resulting in partial mobilization of the passive pressures in the surrounding soil:

$$\Delta r_c = \Delta\sigma_{hdif}\, E \cdot (1/a_c - 1) \cdot r_c) \tag{7.34}$$

The objective is to obtain the value of $\Delta r_c$ in order to obtain the force acting on the geosynthetic (Eq. 7.32) and the settlement (Eq. 7.38) result due to loading ($\Delta\sigma_0$) generated by the construction of the embankment over the column.

The horizontal deformation of the column $\Delta r_c$ and the settlement of the soil $\Delta h_s$ (accepted as equal to the settlement of the column $\Delta h_c$) can be calculated by an iterative process using Eq. (7.35) and (7.36). The calculations are based on the approach proposed by Ghionna and Jamiolkowsky (1981). In the iterative process, one must determine the value of $\Delta\sigma_{vs}$, and then the value of $\Delta r_c$ using Eq. (7.36).

$$\left\{\frac{\Delta\sigma_{v,s}}{E_{oed,s}} - \frac{2}{E^*}\cdot\frac{v_s}{1-v_s}\left[\begin{array}{l} K_{a,c}\left(\dfrac{1}{a_c}\cdot\Delta\sigma_0 - \dfrac{1-a_c}{a_c}\cdot\Delta\sigma_{v,s} + \sigma_{v,0,c}\right) - \\ K_{0,s}\cdot\Delta\sigma_{v,s} - K_{0,s}^*\cdot\sigma_{v,0,s} + \dfrac{(r_{geo}-r_c)\cdot J}{r_{geo}^2} - \dfrac{\Delta r_c\cdot J}{r_{geo}^2} \end{array}\right]\right\}\cdot h_c$$
$$= \left[1 - \frac{r_c^2}{(r_c+\Delta r_c)^2}\right]\cdot h_c \tag{7.35}$$

$$\Delta r_c = \frac{K_{a,c}\cdot\left(\dfrac{1}{a_c}\cdot\Delta\sigma_0 - \dfrac{1-a_c}{a_c}\cdot\Delta\sigma_{v,s} + \sigma_{v,0,c}\right) - K_{0,s}\cdot\Delta\sigma_{v,s} - K_{0,s}^*\cdot\sigma_{v,0,s} + \dfrac{(r_{geo}-r_c)\cdot J}{r_{geo}^2}}{\dfrac{E^*}{(1/a_c-1)\cdot r_c} + \dfrac{J}{r_{geo}^2}} \tag{7.36}$$

where:
$a_c = A_c/A$ (ratio of the areas);
$\Delta\sigma_0$ – increase of vertical stress (embankment over columns);
$\sigma_{voc}$ – initial vertical stress (without surcharge) of the column soil;
$\sigma_{vos}$ – initial vertical stress (without surcharge) of the surrounding soil;
$K_{ac}$ – active pressures coefficient generated by the material of the column.
The value of $E^*$ is given by:

$$E^* = \left(\frac{1}{1-v_s} + \frac{1}{1+v_s}\cdot\frac{1}{a_c}\right)\cdot\frac{(1+v_s)\cdot(1-2v_s)}{(1-v_s)}\cdot E_{oed,s} \tag{7.37}$$

where $v_s$ is the Poisson ratio of the soil and $E_{oeds}$ is the oedometer (constrained) modulus of the soil.
The settlement of the column-soil system is given by:

$$\Delta h_s = \Delta h_c = \left(1 - \frac{r_c^2}{(r_c+\Delta r_c)^2}\right)\cdot h_c \tag{7.38}$$

where $h_c$ is the height of the column and $r_c$ is the initial radius of the column.

It is recommended to update the values $r_c$ and $h_c$ based on values from $\Delta r_c$ and $\Delta h_s$ (or $\Delta h_c$ during the iterations). For preliminary calculations, it is recommended to use the oedometer modulus of the soil as a constant value. For a more precise calculation, one should consider the modulus of the soil dependency on the stress level (see Equation 7.39), and its variation is represented by the equation:

$$E_{oed,s} = E_{oed,s,ref} \cdot \left( \frac{P^*}{P_{ref}} \right)^m \tag{7.39}$$

where:

$E_{oedsref}$ – reference oedometer modulus (modulus obtained for stress $P_{ref}$);
$P_{ref}$ – reference stress;
$E_{oeds}$ – oedometer modulus for a given stress;
$P^*$ – active earth pressure;
m – exponent.

For practical applications, one may use the value $P^*$ (Kempfert and Gebreselassie, 2006), given by the following equation:

$$P^* = \frac{(\sigma_2^* + \sigma_1^*)}{2} \tag{7.40}$$

which:

$$\sigma_{1,2}^* = \frac{1}{2} \cdot \{(\Delta\sigma_{v,s} + \sigma_{v,0,s}) + [K_{0,s} \cdot \Delta\sigma_{v,s} + K_{0,s}^* \cdot \sigma_{v,0,s} + \Delta\sigma_{h,diff}]\} + c_s \cdot \cot \phi_s \tag{7.41}$$

where $c_s$ is the soil cohesion and $\phi_s$ is the friction angle of the soil. In this case, $\sigma_1^*$ and $\sigma_2^*$ are the stresses before and after loading, respectively.

A parametric study by Alexiew, Brokemper and Lothspeich (2005) for a typical example of embankment on soft soil, shown in Figure 7.22, illustrates the application of the equations above. In this study, there was a variation of the following parameters: modulus J of the geotextile, between 1000 kN/m and 4000 kN/m; height of the embankment, between 6 m and 14 m; column spacing, in terms of $a_c$ replacement ratio between 10% and 20%. The results of this study for the oedometer modulus value of the soil equal to 500 kPa are given in Figure 7.23.

Almeida et al. (2013) compared Raithel (1999)'s analytical solution with two-dimensional finite element analyses and observed good agreement. Parametric studies showed that the influence of the geosynthetic encasement on settlement improvement is proportionally greater in shallow soft soil layers.

## 7.3.4 Case histories for applications of embankments over encased granular columns

Encased sand columns were first used in South America on a highway (Mello et al., 2008) near the city of São José dos Campos, 100 km from São Paulo. The subsoil at that location is composed of two soft clay layers separated by a silty sand layer. The columns were installed using Franki pile driving equipment with a closed end. After installation of the encasing, the sand was placed inside the geosynthetic and the

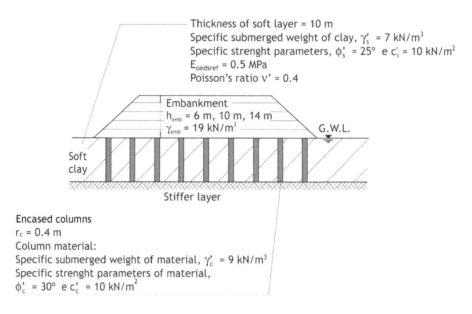

*Figure 7.22* Outline of the embankment analysis (Alexiew, Brokemper and Lothspeich, 2005).

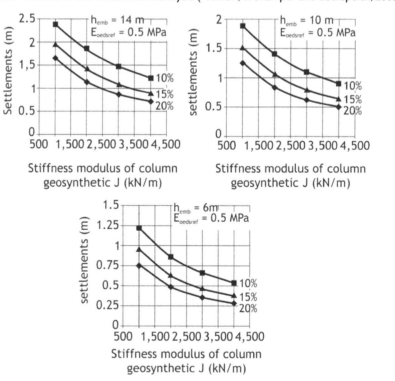

*Figure 7.23* Settlements x encasing modulus for $E_{oedsref} = 0.5$ MPa (Alexiew, Brokemper and Lothspeich, 2005).

*Figure 7.24* Execution details of encased columns.

*Table 7.3* Summary of the column characteristics and results (Mello et al., 2008).

| Characteristics | Values |
| --- | --- |
| Diameter of columns | 0.70 m |
| Geotextile used for encasing | ultimate stress of 130 kN/m and stiffness of 2,000 kN/m |
| Length of columns | ≈10 m |
| Spacing | 1.85 and 2.2 m |
| Measured settlements | 100 mm |
| Stabilization period after initial readings | 6 months |

*Table 7.4* Characteristics of the loaded area and columns (Alexiew and Moormann, 2009).

| Characteristics | Values |
| --- | --- |
| Diameter of columns | 0.78 m |
| Length of columns | 10 m–12 m |
| Spacing | 2 m × 2 m |
| Geotextile used for encasing | Ringtrac 100/250 and 100/275 |

tube was removed using a vibrating hammer. Figure 7.24 shows the column in the final stages of execution, and Table 7.3 summarizes the characteristics of the columns and some monitoring results.

For the construction of part of the CSA coke stockyard (Alexiew and Moormann, 2009) encased sand columns were also used. In the area where this technique was used, the local subsoil consisted of very soft and compressible clay layers. The characteristics of the columns used at this location are summarized in Table 7.4.

To assess the effectiveness of the encased columns in relation to stone column foundations, the results of field tests are presented Figure 7.25 (Kempfert, 2003) where

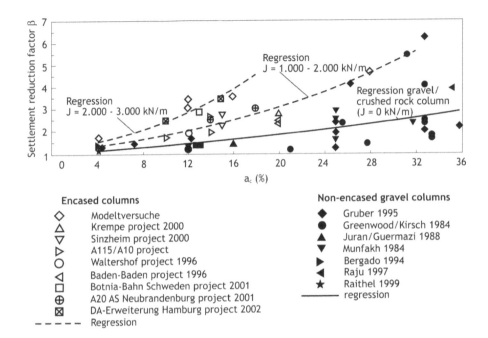

*Figure 7.25* Soil improvement factors depending on area replacement ratio (Kempfert, 2003).

the improvement ratio β is plotted against the area ratio and it is clearly observed that encased columns present greater β values than stone columns.

## 7.4   FINAL REMARKS

This chapter described three construction methods and design procedures for embankments over pile-like elements. Smaller settlements and faster execution are the main advantages of these construction methods compared to traditional methods. The three methods described here have been used in very soft soils and have shown better performance for moderately to high embankments.

The use of embankments with improved geotechnical strength properties and adequate compaction improves the overall performance of such construction techniques. The same may be said about the use of geosynthetics at the base of embankments over granular columns. The successful application of these techniques requires careful geotechnical design, with details for each component (column, cap, fill, geosynthetic) and the interface between them, in addition to careful execution on site.

Considering the diversity of materials employed in the techniques described herein (fill, soft soil, geosynthetic and pile or column) with different strength and deformability characteristics, numerical methods are recommended, such as Finite Element for complementary analyses. The embankment on reinforced concrete slab and piles technique has been used mainly in cases of very soft and thick layers with tight construction deadlines and when post-construction settlements should be virtually nil.

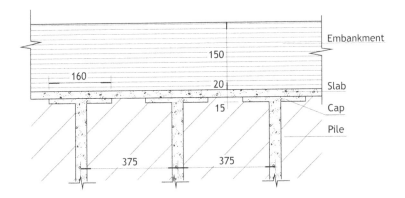

*Figure 7.26* Example of an embankment over concrete slab.

This technique consists in the construction of a slab over a network of piles with caps, as shown in Figure 7.26.

Compared with the geogrid, the slab does not present embankment deformations in the medium and long term and, moreover, the embedment of the piles contributes to a better distribution of horizontal loads on the periphery of the embankment. The main disadvantage of this technique is the high cost.

# Chapter 8

# Monitoring embankments on soft soils

Embankments on soft soil are monitored to fulfill objectives that include assessment of the design assumptions; planning of the field work, especially in terms of loading and unloading stages; and to ensure the structural integrity of the neighboring constructions. In order to achieve the above objectives some important criteria must be met:

- The magnitudes of each type of measurement as well as the range of expected variations must be known in advance;
- The analyses should be performed immediately after the readings, to provide adequate timing concerning the field works;
- The plan of instrumentation should inform how and where the instruments will be installed (e.g., location and depth), the recommended reading frequencies and how the measures will be taken, the warning values and the decisions associated with such values. It should also inform how often analyses reports should be presented;
- The instruments must be located by coordinates and altimetry. The instrumentation should be installed, whenever possible, close to places where previous site investigations have been performed.

It is out of the scope of this book to describe each type of instruments used, with their advantages and disadvantages, for which the reader should refer to the relevant technical literature (e.g., Dunnicliff 1998). Instruments generally used for monitoring the behavior of embankments on soft clays in general are shown schematically in Figure 8.1a the main instruments used for stability control, and in Figure 8.1b the instruments used for settlement control. The instruments presented in Figure 8.1 are described in the items below. Each instrumentation plan should be adapted to the specific type of construction method or ground treatment adopted.

## 8.1  MONITORING VERTICAL DISPLACEMENTS

### 8.1.1  Settlement plates

The settlement plate, used to measure the vertical displacements, is the simplest instrument used in any monitoring plan. It is usually composed of a metallic square plate screwed to rods, as shown in Figure 8.2, thus allowing its extension with the elevation of the embankment. The PVC tube surrounding the rod minimizes the rod-fill

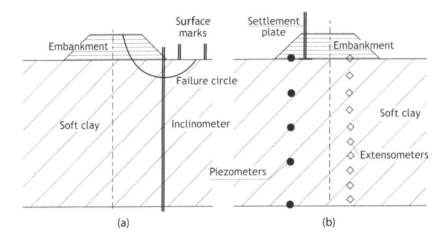

Figure 8.1 Schematic outline of section of monitored embankments over soft soils.

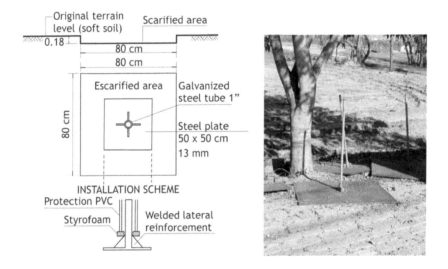

Figure 8.2 Detail of a settlement plate.

material friction. The presence of a benchmark near the embankment is fundamental for settlement plates monitoring.

To avoid damage to the plates, a protective enclosure is quite commonly used around them and it is removed during the elevation of the embankment to allow careful compaction around the plate (Figure 8.3). The integrity of the instruments must be garanteed, although accidents with instrumentation are quite common, thus more plates are sometimes installed to ensure adequate overall monitoring interpretation.

Settlement plates are simple to use and easy to install, thus they should be installed at the very early stages of the fill placement, so that records of settlements are not lost

*Figure 8.3* Detail of the settlement plate in the field (note the careful compacting around the plate) and the fence protection.

during this stage. The location of the plates must be such as to allow comparison between the reading result and design calculations. Thus, it is recommended that settlement plates are installed near the boreholes and away from the embankment edge areas, where analysis is more complex.

The frequency of readings depends on the execution schedule of the embankment and the rate of fill compaction. Usually, during construction of an embankment, readings are carried out twice a week, then once a week after the end of construction. The team hired to do altimetry readings should also inform the thicknesses of the fill where plates are installed for each reading. The schedule of the readings can change according to the characteristics of embankment like height and characteristics of soft soil foundation such as the coefficient of consolidation.

## 8.1.2 Depth extensometers

While settlement plates measure the total surface settlements below the embankment, depth extensometers provide in-depth measurements of settlements associated with the sub layers with distinct geotechnical characteristics. The extensometers are installed inside the soft clay layer, as shown in Figure 8.1b and are most commonly used in large projects.

The magnetic extensometer (Figure 8.4) is the most commonly used for in-depth settlement measures and consists of installing a tube along the soft layer, through the anchoring in a fixed unmovable reference. Magnetic rings (spider type) are installed around the tube and anchored in the soil around it, to enable its displacement together with the soil. A probe coupled to a graduated tape is introduced in the tube with a beeping device at the tip that is activated when it passes the magnetic ring. The readings are taken according to the fixed reference at the bottom tip of the tube.

A number of different types of depth extensometers, based on a number of different principles may be used, some of them associated with automated data logging, for instance but it is out of the scope of this book to describe them.

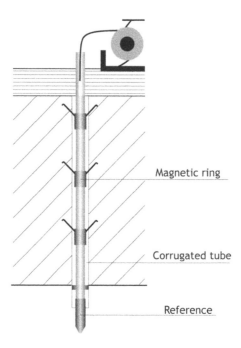

*Figure 8.4* Detail of a magnetic extensometer.

### 8.1.3   Settlement Profiler

The settlement profiler (Palmeira and Ortigão, 1981; Borba, 2007) provides a continuous settlement profile along a horizontal line, and this is the main advantage when compared to the settlement plates, which provide individual settlements. Figure 8.5 compares results for both instruments.

With the settlement profiler a tube is installed at the base of the embankment, very similar to the inclinometer (described ahead), with a rope inside to pull the probe.

Difficulty may arise when using this type of instrument if the settlements are large, because it may be difficult to pass the probe through the tube. In this case, the settlement profiler becomes inoperative, which may be prevented by performing settlement calculations in advance. The profilometer probe may be analogous to the inclinometer probe.

An advantage of the settlement profiler with respect to settlement plates is that there are no measuring rods in the embankment, since these usually interfere with the movement of heavy earthmoving equipment.

### 8.2   MEASUREMENT OF HORIZONTAL DISPLACEMENTS

The inclinometer is the instrument used for measuring the horizontal displacement along a vertical distance, through the measurement of the deviation of the tube in

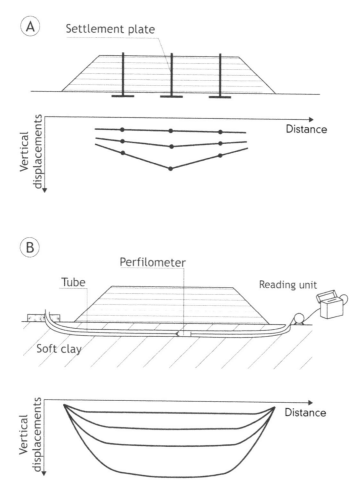

*Figure 8.5* Settlement measurements: (A) from the settlement plates; (B) from the settlement profiler (Ortigão and Almeida, 1988 – DNER-PRO 381/98).

relation to the vertical. The inclinometer (guide tube) installed in the soil through a stiffer layer (Figure 8.6A), has grooves along its length (Figure 8.6B) and it may be of metal or PVC. A probe (probe – Figure 8.6C) with retractable wheels is inserted into this tube, and the wheels guarantee the alignment throughout its passage inside the pipe. The grooves of the tube also serve to indicate the direction of the readings in relation to the construction (Figure 8.6B). For embankments on soft soil, the inclinometer tube should be installed so that the grooves are perpendicular to the foot of the embankment, assuring that the greatest displacement is reading in the same direction (AA) as a pair of alignment grooves. Regardless, readings should be taken in the perpendicular direction (BB) and, if necessary, the vector resultant of the displacement between directions AA and BB is then calculated.

Since horizontal displacements can be very high, it is recommended that the integrity of the tube be verified before each reading. This is done by lowering a dummy

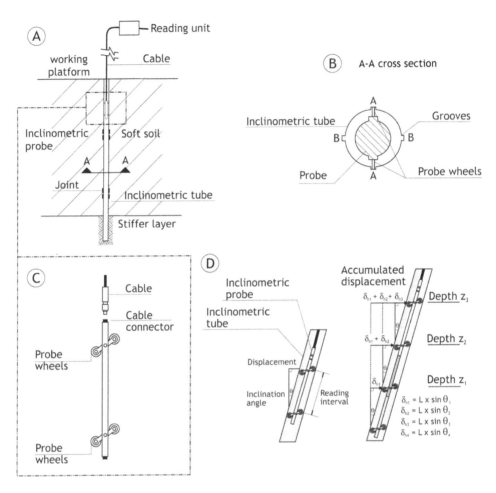

*Figure 8.6* Schematic detail of an inclinometer: (A) and (B) inclinometer tube and inclinometer probe; (C) inclinometer probe; (D) detail of readings.

probe before the readings to avoid losing the actual probe. Figure 8.6D shows the calculation of the accumulated deviations measured by an inclinometer probe through the tube, which makes the calculation of the accumulated displacements possible. Readings are performed at constant intervals (L = 0.5 m, 1 m etc.), in the ascending direction.

## 8.3  MEASUREMENTS OF PORE PRESSURES

Several types of piezometers may be used for pore pressure measurements. The Casagrande piezometer (open tip – Figure 8.7A) has a long response time but may be suitable for long term measurements. The filter, located at the bottom of the piezometer (installation depth), is comprised of a perforated PVC tube wrapped in geotextile to minimize clogging.

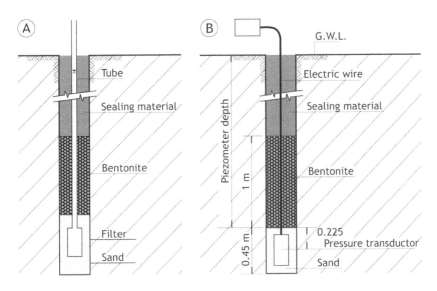

*Figure 8.7* Piezometer scheme outline: (A) Casagrande; (B) electric or vibrating wire.

Electric or vibrating wire piezometers (Figure 8.7B), although more expensive, have a shorter response time than Casagrande's piezometer, do not interfere with the compaction process and can also measure negative pore pressures that occur during vacuum preloading. On the other hand, it is possible to do permeability tests *in situ* when using Casagrande's piezometer, which is not possible with the electrical piezometer.

## 8.4   MONITORING OF THE TENSILE FORCES IN GEOSYNTHETIC REINFORCEMENTS

Almeida et al. (2007) describe the instrumentation of a piled embankment with a geogrids platform. The tensile forces measurement on the geogrid was taken by fixing three sensors using a connector comprised of a steel rod attached to a sphere, with its interior moving freely in all directions. This element was used to prevent a moment in the system. The sensors were protected by a hard PVC tube after installation. This system was also used (reinforced soil wall), as presented in Riccio (2007).

Magnani, Almeida and Ehrlich (2009) describe the instrumentation of a reinforced embankment on soft soils, where load cells were installed in the reinforcement to measure tensile stress. These load cells designed to endure the construction efforts of earthmoving activities as well as to overcome the relaxation effects on geosynthetics. These cells were connected in a 1.5 m wide band of geosynthetic material, so that the geosynthetic fibers were aligned and then secured with epoxy resin.

The instrumentation described by Almeida et al. (2007) and Magnani, Almeida and Ehrlich (2009) was developed in house and has been used in quite a few studies.

Fiber optic instrumentation is increasingly for monitoring geosynthetic tensile forces (Briançon et al. 2010).

## 8.5   INTERPRETATION OF MONITORING RESULTS

Analyses of settlements s(t) and pore pressure $\Delta u(t)$ versus time can provide important design information. Examples of there are field coefficient of consolidation and final settlement values $s_\infty$, Chart 8.1 shows some calculation methods proposed by several authors and the parameters obtained.

### 8.5.1   Asaoka's method (1978)

Asaoka (1978) proposed a simple method for the interpretation of the results of settlement measurements over time, to obtain vertical consolidation coefficients and prediction of final settlements. The method is widely used but some authors have strong arguments against the method as discussed below. The procedures of using Asaoka's method (1978) are:

1   draw the curve s × t and define the value for constant $\Delta t$;
2   get the values of equally spaced s by $\Delta t$ (Figure 8.8A) and plot them in diagram $s_i \times s_{i-1}$ (Figure 8.8B);
3   adjust a straight line through the points and obtain the inclination $\beta_1$ and calculate $c_v$, $c_h$ by Eqs. (8.1) (vertical drainage) and (8.2) ( radial drainage);
4   draw a straight line of 45°, $s_i = s_{i-1}$, and obtain the final settlement $s_\infty$.

Values of $c_v$ and $c_h$ are calculated by:

$$c_v = -\frac{5}{12} \cdot h_d^2 \cdot \frac{\ln \beta_1}{\Delta t} \tag{8.1}$$

$$c_h = -\frac{F(n)}{8} \cdot d_e^2 \cdot \frac{\ln \beta_1}{\Delta t} \tag{8.2}$$

Further details about how to obtain the values of F(n) (Eq. 5.8) and $d_e$ (Eqs. 5.11 and 5.13) are presented in Chapter 5.

Chart 8.1   Calculation methods for evaluating performance of embankments on soft soils.

| Calculation methods | Necessary data for calculation | Parameters obtained |
|---|---|---|
| Ellstein (1971) | s(t) | $s_\infty, c_v$ |
| Long and Carey (1978) | s(t) | $s_\infty, c_h$ |
| Tan (1971) | s(t) | $s_\infty$ |
| Asaoka (1978)* | s(t) | $s_\infty, c_v, c_h$ |
| Scott (1961) | s(t) | $c_v, c_h$ |
| Escario and Uriel (1961) | s(t) and $s_\infty$ | $c_h$ |
| Orleach (1983)* | $\Delta u(t)$ | $c_v, c_h$ |

(*) presented ahead.

For combined radial and vertical drainage, a value of $c_v$ should be assigned and $c_h$ should be determined, using the following equations:

$$-\frac{\ln \beta_1}{\Delta t} = \frac{8c_h}{d_e^2 \cdot F(n)} + \frac{\pi^2 c_v}{4h_d^2} \tag{8.3}$$

$$c_h = \frac{-d_e^2 \cdot F(n)}{8} \cdot \left(\frac{\ln \beta_1}{\Delta t} + \frac{\pi^2 c_v}{4h_d^2}\right) \tag{8.4}$$

Time intervals ($\Delta t$) should be between 30 and 90 days and at least three intervals are necessary for settlement estimates and $c_v$ or $c_h$ field estimates, i.e. only after this period is it possible to obtain results that would lead to a decision by the construction team.

## 8.5.2 Pore pressure analysis

Pore pressure data may be interpreted according to what is proposed by Orleach (1983). Values of $c_v$ and $c_h$ are calculated by:

$$c_v = -\frac{4h_d^2 \alpha_1}{\pi^2} \tag{8.5}$$

$$c_h = -\frac{F(n)d_e^2 \alpha_1}{8} \tag{8.6}$$

where $\alpha_1$ may be obtained by means of the $\log(\Delta u)$ diagram versus t, according to Figure 8.9.

Since Eq. (8.6) was developed for pure radial drainage, the piezometers used for the analyses must be carefully chosen. For instance, in case of vertical drains in which top and botton layers are draining interfaces the piezometers chosen should be those located closer to the middle of the layer.

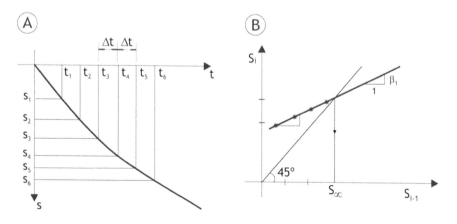

Figure 8.8 Graphic construction of Asaoka's method (1978): (A) time × settlement curve; (B) adjusted line.

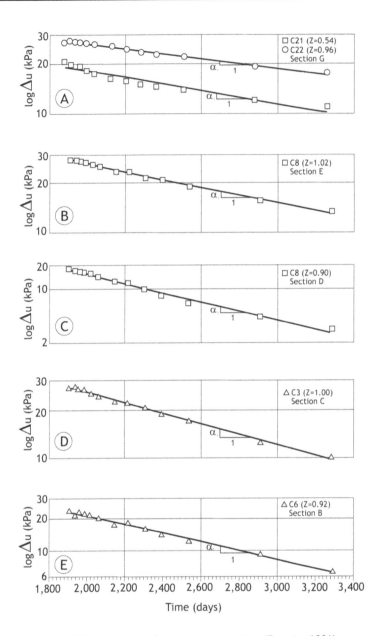

*Figure 8.9* Log curves of pore pressure × time (Ferreira, 1991).

## 8.5.3  Discussion on obtaining $c_v$ and $c_h$ from monitoring data

Table 8.1 presents results of coefficients of consolidation obtained from in situ and laboratory tests compared to field monitoring data for two Rio de Janeiro soft clay deposits.

Results for Sarapuí clay shown in Table 8.1, present general consistency of results for $c_h$, but not for $c_v$. Values of $c_v$ obtained from settlement data are greater than

*Table 8.1* Values of $c_v$ and $c_h$ from laboratory and field tests and from monitoring.

*Sarapuí Clay (Almeida and Ferreira, 1992)*

| Calculation methods or direct measure | Depth (m) | $c_v$ (10$^{-8}$ m$^2$/s) | $c_h$ (10$^{-8}$ m$^2$/s) |
|---|---|---|---|
| Oedometer test, Method of Taylor (Coutinho, 1976) | 5–6 | 1.2 | 2.4 |
| Piezocone, Houlsby and Teh (1988) (Danziger, 1990) | 2.2–8.2 | 1.6–4.4 | 3.1–8.7 |
| Field instrumentation, settlement plates Asaoka (1978) (Schmidt, 1992) | whole layer | 17.8 | 3.1–4.4 |
| Field instrumentation, magnetic extensometers, Asaoka (1978) (Almeida et al., 1989) | whole layer | 22.6 | 4.2–8.1 |
| Field instrumentation, piezometers Casagrande, Orleach (1983) (Ferreira, 1991) | 3.3-8.3 | 2.2–4.5 | 1.2–2.8 |

*Clay from Barra da Tijuca (Almeida et al., 2001)*

| Calculation methods or direct measure | Variation range of $c_h$ (10$^{-8}$ m$^2$/s) | $c_h$ (average) (10$^{-8}$ m$^2$/s) |
|---|---|---|
| Field instrumentation, settlement plates Asaoka (1978) | 3.7–10.5 | 6.8 |
| Piezocone (Houlsby and Teh, 1988) | 2.4–13.7 | 8.2 |
| Oedometer tests with radial drainage | 3.6–6.8 | 5.0 |

obtained with piezometer, which seems to be a reflex of secondary compression on the settlement.

It is noted that for the Barra da Tijuca clay (Table 8.1), the general consistency of the values is probably due to the use of geodrains for the acceleration of settlements. in two soft soil deposits in Rio de Janeiro.

When secondary compression settlements are of considerable magnitude and occur in parallel with primary consolidation settlements, results of $c_v$ provided by Asaoka's method (1978) may be quite different from in situ and laboratory tests (Almeida and Ferreira, 1992; Schmidt, 1992; Pinto, 2001). This is because the value of $\beta_1$ (Figure 8.8D) is affected by the secondary compression settlements, thus values of $c_v$ are greater than the measured valued in laboratory. The same happens with $c_h$, but to a much lower degree because primary consolidation is faster when vertical drains are present and the secondary occurs mostly after the primary consolidation.

Low values of applied stress ratio in relation to the initial stress ($\Delta\sigma'_{vf}/\Delta\sigma'_{vo}$) can lead to errors in the values of $c_v$ obtained with the Asaoka's method (Schmidt,1992) since in this case, the secondary compression is relatively important. This is because the lower the value of the increment ratio $\Delta\sigma_v/\sigma_v$, the greater the influence of secondary compression on the settlement curve (Leonards and Girault, 1961).

Almeida et al. (1993) obtained similar $c_v$ values from in situ and laboratory tests, compared with the field values obtained with the Asaoka method (1978), as in this case the high $\Delta\sigma'_v$ (embankment height around 24 m), led to hardly any secondary compression.

Therefore, the differences in predicted versus consolidation coefficients values obtained through different methods are due to factors such as:

- in laboratory analysis is one-dimensional, but in the field boundary conditions are different;
- existing sand lenses make difficult a comparison field versus small scale laboratory tests;
- observed field secondary compression influences the analysis of the results.

### 8.5.4   Stability of embankment by horizontal displacements analysis

The stability of an embankment on soft soil may be qualitatively evaluated based on inclinometer measurements. From the inclinometer readings, it is possible to calculate distortions along the inclinometric tube. The distortion d is the arc tangent of the straight line connecting two consecutive points of the horizontal displacement curve against the depth, and calculated by:

$$d = arctg\left(\frac{\delta_{h1} - \delta_{h2}}{z_1 - z_2}\right) \tag{8.7}$$

where $\delta_{h1}$ and $\delta_{h2}$ are the horizontal displacements (Figure 8.6D) in depths $z_1$ and $z_2$ respectively. Figure 8.10A shows maximum horizontal displacements over time at a given depth of the inclinometer. It is worthy to note that the depth of these values may change over time with the consolidation process. Figure 8.10B shows the distortion profiles measured over time. It is observed that at a depth of 5 m, distortion is maximum, indicating the yield of the clay at this depth.

The distortion rate can then be calculated for the given depth as:

$$v_d = \frac{\Delta d}{\Delta t}(\%/day) \tag{8.8}$$

Almeida, Oliveira and Spotti (2000) recommend that:

- Precautions are necessary for $v_d > 1.5\%$ per day. This can include the interrupting of loading;
- Special attention is required, for $v_d$ between 0.5% and 1.5% per day, since the yield has initiated but has not been completely propagated; it is recommended increasing the frequency of readings;
- There are no major concerns at first for $v_d < 0,5\%$ per day. It is advised the continuous monitoring, until stabilization can be verified.

It must be noted that the installation of an inclinometer at a specific area does not indicate that the whole construction will present a similar distortion rate. This is due to stratigraphy and loading rate variability, which can be different throughout the construction.

Sandroni, Lacerda and Brandt (2004) proposed a method for safety evaluation of embankments on soft soils where volumes are estimated from vertical ($V_v$) and horizontal ($V_h$) displacements, which should be obtained since the beginning of the

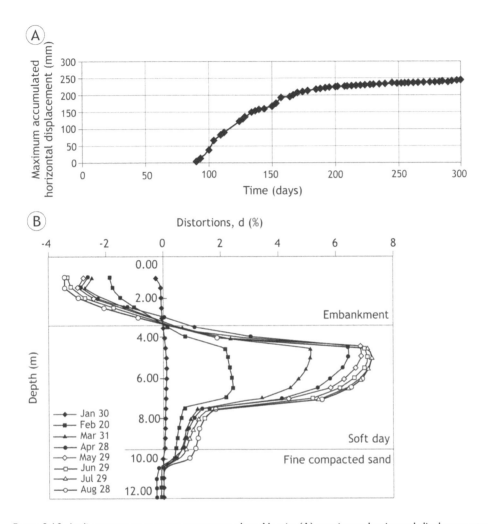

*Figure 8.10* Inclinometer at sewage treatment plant Alegria: (A) maximum horizontal displacements; (B) distortion profiles over time (Almeida, Oliveira and Spotti, 2000).

construction (Figure 8.11). The procedure takes into consideration the fact that volumes $V_v$ and $V_h$ are similar, considering plane stress and undrained failure conditions. When there is tendency to failure, the ratio $V_v/V_h$ has a sharp drop, tending to unity; when there is loading interruption (loading by stages, for example), $V_v/V_h$ tends to increase with consolidation.

Magnani et al. (2008) applied these two procedures in embankments with failures and observed good agreement for the two methods described above.

### 8.5.5 *In situ* compression curves

It is possible to obtain the compression curve of a sublayer, as indicated in Figure 8.12 based on pore pressures and extensometers measurements.

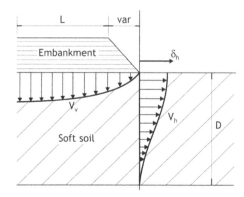

*Figure 8.11*   Details of estimated volumes from monitoring (Sandroni, Lacerda and Brandt, 2004).

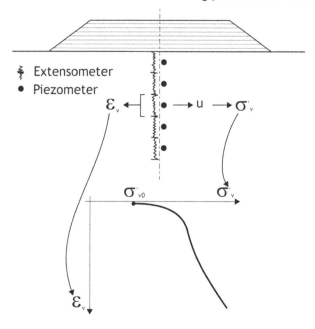

*Figure 8.12*   Detail of the compression curve of a clay sublayer (adapted from Leroueil, 1997).

Marques (2001) monitored an embankment where vacuum preloading was applied. The compression curve was obtained from the pore pressure and deformation measures of two layers (Figure 8.13). Figure 8.14 shows the comparison of compression curves *in situ* with the compression curves from conventional oedometer tests performed in samples collected at the same depth.

The effective vertical stresses at the limit state *in situ* (overconsolidation stress) were lower than those obtained in oedometer tests. The $C_c$ value obtained in oedometer tests was also inferior to those obtained *in situ*. This behavior was also observed by Kabbaj, Tavenas and Leroueil (1988) and by Tavenas and Leroueil (1987), for clays of

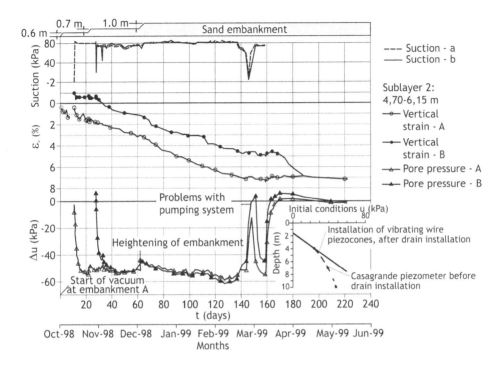

*Figure 8.13* Pore pressure and vertical deformation measurements by vacuum preloading (Marques, 2001).

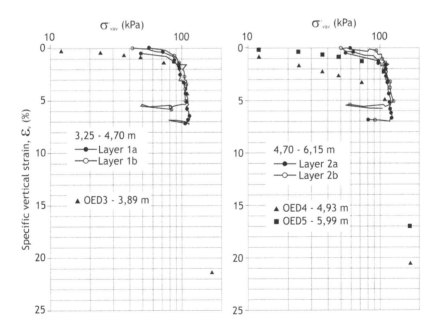

*Figure 8.14* Compression curves in field using vacuum preloading and in laboratory using oedometer tests (Marques, 2001).

Eastern Canada. The values of the effective vertical stress at the limit state of these clays, obtained by compression curves at the end of the primary (EOP – end of primary) or 24 hours in the laboratory were higher than the overconsolidation stresses obtained *in situ*.

The differences between laboratory and field compression curves are caused by several factors. In Figure 8.14, the effect of the temperature difference (field, 7°C; laboratory, 20°C) was superseeded by the effect of low strain rate in the field, when compared to laboratory results. In Brazil, average soil temperatures are approximately 20°C; therefore the differences are more associated to the different strain rates (field that is approximately $10^{-9}$ s$^{-1}$ to $10^{-12}$ s$^{-1}$ and laboratory approximately $10^{-5}$ s$^{-1}$ to $10^{-7}$ s$^{-1}$).

## 8.6   NEW TRENDS IN GEOTECHNICAL INSTRUMENTATION

New instrumentation equipment technologies have appeared in the last decades. Among them a promising one uses optical fiber in the geotechnical field. Some advantages of this technology are: low weight, flexibility, long transmission distance, low material reactivity, electrical isolation and non-sensibility to electromagnetic effects, in addition to providing computerized readings and easy installation and maintenance (Mello, 2013). Many geotechnical sensors are based on this principle – such as piezometers, measuring instruments for deformation, displacement, temperature and pressure. Furthermore, fiber optical inclinometers have also been proposed using a principle quite different from the one presented in section 8.2. Geosynthetics are also being monitored using this technology (Briançon et al., 2010), making it possible to measure deformations and forces in these reinforcement materials. In this case the optical fiber should be attached to the reinforcement, preventing sliding between these two materials.

## 8.7   FINAL REMARKS

Clear goals should be defined in any field monitoring program in terms of limit state conditions and service state conditions. Considering that factors of safety are quite low in embankments on soft soils, it is essential to define the warning ranges.

Different methods are available to evaluate the performance of an embankment on soft soil. As far as consolidation monitoring is concerned the Asaoka (1978)'s method, based on vertical displacements, is quite often used. In the case of significant secondary compression settlements in purely vertical drainage the Asaoka method does not provide satisfactory results for the determination of $c_v$ values.

To determine the values of $c_h$ for mainly radial drainage, the value for the secondary and primary consolidation occurring at the same time is usually small, and the resultant values from $c_h$ are satisfactory.

Sometimes it is not possible to identify the heterogeneity of the layers, thus the coefficient of consolidation used in design may not be verified in the field. Monitoring makes possible to verify the design criteria and propose adjustments during construction.

Monitoring with inclinometers is quite important not just to evaluate the stability during construction but also to assess horizontal displacements affecting nearby structures.

# References

Aas, G. (1965) A study of the effect of vane shape and rate of strain on the measured values of in situ shear strength of clays. In: ISSMGE, *6th International Conference on Soil Mechanics and Foundation Engineering, 1965, Montreal, Canada*, vol. 1, pp. 141–145.

Associação Brasileira de Normas Técnicas (1989) NBR 10905:1989. *Soil – Field vane shear test – Method of test (in Portuguese)*, Brazil, ABNT.

Associação Brasileira de Normas Técnicas (1984) NBR 6508:1984. *Soil particles passing through sieve 4.8 m. Determination of specific mass (in Portuguese)*, Brazil, ABNT.

Associação Brasileira de Normas Técnicas (1996) NBR 13600:1996. *Soil – Determination of organic matter content by igniting at 440°C – Method of test (in Portuguese)*, Brazil, ABNT.

Associação Brasileira de Normas Técnicas (1997) NBR 9820:1997. *Assessment of undisturbed low consistency solid samples from boreholes – Procedure (in Portuguese)*, Brazil, ABNT.

American Society for Testing and Materials (2012) D1587-08:2012. *Standard Practice for Thin Walled Tube Sampling of Soils for Geotechnical Purposes*, United States, ASTM.

American Society for Testing and Materials (2010) D4318-10:2010. *Standard Test Methods for Liquid Limit, Plastic Limit, and Plasticity Index of Soils*, United States, ASTM.

Abusharar, S.W., Zheng, J.-J., Chen, B.-G. & Yin, J.-H. (2009) A simplified method for analysis of a piled embankment reinforced with geosynthetics. *Geotextiles and Geomembranes*, 27, 39–52.

Alencar Jr., J.A. (1984) *Pore pressure analysis related to cone tests carried on Sarapuí soft clay. (in Portuguese)*. Master's Thesis, PUC-Rio, Rio de Janeiro, Brazil.

Alencar JR., J.A., Fraiha Neto, S.H., Saré, A.R. & Mendonça, T.M. (2001) Geotechnical characteristics of some soft to medium clays of Belém city, Pará State (in Portuguese). *In:* COPPE/UFRJ, *Workshop Properties of Brazilian Soft Soils, March 2001*, Rio de Janeiro, Brazil, CD-ROM.

Alexiew, D., Brokemper, D. & Lothspeich, S. (2005) Geotextile encased columns (GEC): load capacity, geotextile selection and pre-design graphs.In: *GSP 131 Contemporary Issues in Foundation Engineering. Geofrontier.*

Alexiew, D., Horgan, G.J. & Brokemper, D. (2003) Geotextile encased columns (GEC): load capacity & geotextile selection. In: BGA, *International Conf. on Foundations, 2003, Dundee, Scotland*, pp. 81–90.

Alexiew, D. & Moormann, C. (2009) Foundation of a coal/coke stockyard on soft soil with geotextile encased columns and horizontal reinforcement. In: ISSMGE, *17th International Conference on Soil Mechanics and Geotechnical Engineering, Alexandria, Egypt*, CD-ROM.

Almeida, M.C.F., Almeida, M.S.S. & Marques, M.E.S. (2008) Numerical analysis of a geogrid reinforced wall on piled embankment. In: IGS, *1st Pan American Geosynthetics Conference & Exhibition, March, 2008, Cancun, Mexico*, CD-ROM.

Almeida, M.S.S. (1981) *Analysis of the behaviour of an embankment on soft clay foundation.* Master's Thesis, University of Cambridge, England.

Almeida, M.S.S. (1982) The undrained behaviour of the Rio de Janeiro clay in the light of critical state theories. *Soils and Rocks*, ABMS, São Paulo, Brazil, 5 (2), 3–24.

Almeida, M.S.S. (1984) *Stage constructed embankments on soft clays.* Doctoral Thesis, Cambridge University, England.

Almeida, M.S.S. (1985) Embankment failure on clay near Rio de Janeiro. *Discussion on Journal of the Geotechnical Engineering Division*, ASCE, 111 (2), 253–256.

Almeida, M.S.S. (1986) Geotechnical properties of a Rio de Janeiro soft clay, in light of Critical State and empirical correlations (in Portuguese). In: ABMS, *8th Brazilian Conference on Soil Mechanics and Foundation Engineering, Porto Alegre, Brazil*, vol. 1, 15–24.

Almeida, M.S.S. (1996) *Embankments Over Soft Soils – from Conception to Performance (in Portuguese).* UFRJ, Rio de Janeiro, Brazil.

Almeida, M.S.S. (1998) Site characterization of a lacustrine very soft Rio de Janeiro organic clay. In: *ISC, 1998, Atlanta, USA*, vol. 2. pp. 961–966.

Almeida, M.S.S. & Marques, M.E.S. (2003) *The behaviour of Sarapuí soft organic clay.* In: T.S. Tan, K.K. Phoon, D.W. Hight & S. Leroueil (eds.) *International Workshop on Characterization and Engineering Properties of Natural Soils, Singapore*, vol. 1, pp. 477–504.

Almeida, M.S.S. & Ferreira, C.A.M. (1992) Field, in situ and laboratory consolidation parameters of a very soft clay. Predictive soil mechanics. *In: Wroth Memorial Symposium, 1992, Oxford*, pp. 73–93.

Almeida, M.S.S. & Ortigão, J.A.R. (1982) Performance and finite element analyses of a trial embankment on soft clay. In: *International Symposium on Numerical Models in Geomechanics, Zurich*, 548–558.

Almeida, M.S.S., Britto, A.M. & Parry, R.H.G. (1986) Numerical modelling of a centrifuged embankment on soft clay. *Canadian Geotechnical Journal*, 23, 103–114.

Almeida, M.S.S., Collet, H.B., Ortigão, J.A. & Terra, B.R.C.S.R. (1989) Settlement analysis of embankment on Rio de Janeiro clay with vertical drains.In: ISSMGE, *Special Volume of Brazilian Contributions to the 12th Int. Conf. on Soil Mech. and Foundation Engineering, Rio de Janeiro, Brazil*, pp. 105–110.

Almeida, M.S.S., Danziger, F.A.B. & Macedo, E.O. (2006) In situ undrained strength from T-Bar tests (in Portuguese). In: ABMS, *13th Brazilian Conference on Soil Mechanics and Foundation Engineering, 2006, Curitiba, Brazil*, vol. 2, pp. 619–624.

Almeida, M.S.S., Danziger, F.A.B., Almeida, M.C.F., Carvalho, S.R.L. & Martins, I.S.M. (1993) Performance of an embankment built on a soft disturbed clay. *In: International Conference on Case Histories in Geotechnical Engineering, 3, 1993, St. Louis, Missouri*, vol. 1. pp. 351–356.

Almeida, M.S.S., Davies, M.C.R. & Parry, R.H.G. (1985) Centrifuge tests of embankments on strengthened and unstrengthened clay foundations. *Géotechnique*, UK, 35(4), 425–441.

Almeida, M.S.S., Ehrlich, M., Spotti, A.P. & Marques, M.E.S. (2007) Embankment supported on piles with biaxial geogrids. *Journal of Geotechnical Engineering – Institution of Civil Engineers (ICE)*, UK, 160 (4), 185–192.

Almeida, M.S.S., Hosseinpour, I., & Riccio, M. (2013) Performance of a geosynthetic-encased column (GEC) in soft ground: numerical and analytical studies. *Geosynthetics International*, 20 (4).

Almeida, M.S.S., Marques, M.E.S. & Baroni, M. (2010) Geotechnical parameters of very soft clays obtained with CPTu compared with other site investigation tools. In: *2nd International Symposium on Cone Penetration Testing, CPT'10, 2010, Huntington Beach, California, USA*, CD-ROM.

Almeida, M.S.S., Martins, I.S.M. & Carvalho, S.R.L. (1995) Constant Rate of and Strain Consolidation of Brazilian Soft Clays. In: *International Congress on Consolidation and Compression of Clayey Soils, Hiroshima, Japan*, vol. 1, pp. 9–14.

Almeida, M.S.S., Marques, M.E.S. & Lima, B.T. (2010) Overview of Brazilian construction practice over soft soils.In: Almeida, M.S.S. (ed.) *Symposium New Techniques for Design and Construction in Soft Clays, Guarujá, Brazil*. São Paulo, Oficina de Textos, pp. 205–225.

Almeida, M.S.S., Oliveira, J.R.M.S. & Spotti, A.P. (2000) Prediction and performance of embankment over soft soils: stability, settlements and numerical analysis (in Portuguese). In: ABMS/NRSP, *Technical Meeting Prevision of Performance x Real Behaviour, 2000, São Paulo*, pp. 69–94.

Almeida, M.S.S., Futai, M.M., Lacerda, W. & Marques, M.E.S. (2008) Laboratory behaviour of Rio de Janeiro soft clays. Part 1: index and compression properties, *Soils and Rocks*, 31, 69–75.

Almeida, M.S.S., Marques, M.E.S., Almeida, M.C.F. & Mendonça, M.B. (2008a) Performance of two "low" piled embankments with geogrids at Rio de Janeiro. In: IGS *1st Pan American Geosynthetics Conference & Exhibition*, 2008, Cancun, Mexico, CD-ROM.

Almeida, M.S.S., Marques, M.E.S., Lima, B.T.E. & Alvez, F. (2008b) Failure of a reinforced embankment over an extremely soft peat clay layer. *In: European Conference on Geosynthetics-Eurogeo, 4, 2008, Edinburgh*, vol. 1. pp. 1–8.

Almeida, M.S.S., Marques, M.E.S., Miranda, T.C. & Nascimento, C.M.C. (2008c) Lowland reclamation in urban areas. In: TCIU-ISSMGE, *Workshop on Geotechnical Infrastructure for Mega Cities and New Capitals, Aug. 2008, Búzios, Rio de Janeiro, Brazil*.

Almeida, M.S.S., Santa Maria, P.E.L., Martins, I.S.M., Spotti, A.P. & Coelho, L.B.M. (2001) Consolidation of a very soft clay with vertical drains. *Géotechnique*, UK, 50 (6), 633–643.

Almeida, M.S.S., Tejada, F.M.S.F., Santos, H.C. & Muller, H. (1999) Construcción del meulle de múltiplo uso del Puerto de Sepetiba. In: *11th Panamerican Conference on Soil Mechanics and Geotechnical Engineering, Foz do Iguaçu, Brazil*, vol. 3, pp. 1091–1096.

Antunes Filho (1996) *Numerical analysis of Juturnaíba embankment on soft organic clay (in Portuguese)*. Master's Thesis, COPPE/UFRJ, Rio de Janeiro, Brazil.

Asaoka, A. (1978) Observational procedure of settlement prediction. *Soils and Foundations*, 18 (4), 67–101.

Atkinson, J.H. & Bransby, P.L. (1978) *Mechanics of Soils. An Introduction of Critical State Soil Mechanics*. McGraw-Hill, New York, United States. 578p.

Atkinson, M.S. & Eldred, P. (1981) Consolidation of soil using vertical drains. *Géotechnique*, UK, 31 (1), 33–43.

Azzouz, A.S., Baligh, M.M. & Ladd, C.C. (1983) Corrected field vane strength for embankment design. *Journal of Geotechnical Engineering, ASCE*, 109 (5), 730–734.

Balaam, N.P. & Booker, J.R. (1981) Analysis of rigid rafts supported by granular piles. *Int. Journal for Numerical and Analytical Methods in Geomechanics*, 5, 379–403.

Baptista, H.M. & Sayão, A.S.J.F. (1998) Geotechnical characteristics of the soft soil deposit at Enseada do Cabrito, Salvador, Bahia, Brazil (in Portuguese). In: ABMS, *11th Brazilian Conference on Soil Mechanics and Foundation Engineering, 1998, Brasília, Brazil*, vol. 2. pp. 911–916.

Barksdale, R.D. & Bachus, R.C. (1983) *Design and construction of stone columns*. National Technical Information Service, Springfield, Virginia, USA. Report number: FHWA/RD-83/026.

Baroni, M. & Almeida, M.S.S. (2012) In situ and laboratory parameters of extremely soft organic clay deposits. In: *International Conference on Geotechnical and Geophysical Site Characterization (ISC'4), Porto de Galinhas, Brazil*, CD-ROM.

Baroni, M. (2010) *Geotechnical investigation of soft soils of Barra da Tijuca, with emphasis of in situ tests (in Portuguese)*. Master's Thesis, COPPE/UFRJ, Rio de Janeiro, Brazil.

Barron, R.A. (1948) Consolidation of fine-grained soils by drain wells. *Journal of the Soil Mechanics and Foundation Division, ASCE*, 73 (6), 811–835.

Baumann, V. & Bauer, G.E.A. (1974) The performance of foundations on various soils stabilized by the Vibro-Compaction Method. *Canadian Geotechnical Journal*, 11(4), 509–530.

Bedeschi, M.V.R. (2004) *Settlements of an embankments built over very soft deposit with vertical drains at Barra da Tijuca, Rio de Janeiro (in Portuguese)*. Master's Thesis, COPPE/UFRJ, Rio de Janeiro, Brazil.

Bergado, D.T., Asakami, H., Alfaro, M.C. & Balasubramaniam, A.S. (1991) Smear effects of vertical drains on soft Bangkok clay. *Journal of the Soil Mechanics and Foundation Division, ASCE*, 117 (10), 1509–1529.

Bergado, D.T., Chai, J.C., Alfaro, M.C. & Balasubramaniam A.S. (1994) *Improvement techniques on soft ground in subsiding and lowland environment*. Netherlands: Balkema.

Bezerra, R.L. (1993) *Tests with COPPE's second generation of piezocone on assessment of geotechnical characteristics of a clayey deposit of Santos Lowlands (in Portuguese)*. Qualify Seminar for Doctoral Thesis, COPPE/UFRJ, Rio de Janeiro, Brazil.

Bezerra, R.L. (1996) *Development of third generation of piezocone at COPPE/UFRJ (in Portuguese)*. Doctoral Thesis, COPPE/UFRJ, Rio de Janeiro, Brazil.

Bjerrum, L. (1972) Embankments on soft ground. In: ASCE, *Specialty Conference on Earth and Earth- Supported Structures, West Lafayette, United States*, 1972, vol. 2, pp. 1–4.

Bjerrum, L. (1973) Problems of soil mechanics and construction on soft clays. In: ISSMGE, *8th International Conference on Soil Mechanics and Foundation Engineering*, 1973, *Moscow, Russia*, vol. 3, pp. 111–159.

Blanc, M., Rault, G., Thorel, L. & Almeida, M. S. S. (2013) Centrifuge investigation of load transfer mechanisms in a granular mattress above a rigid inclusions network. *Geotextiles and Geomembranes*, Elsevier, 36, 92–105.

Bonaparte, R. & Christopher, B.R. (1987) Design and construction of reinforced embankments over weak foundations. In: *Symposium of Reinforced Layered Systems*, TRB.

Borba, A.M. (2007) *Analysis of the performance of experimental embankment at Pan American Vila (in Portuguese)*. Master's Thesis, COPPE/UFRJ, Rio de Janeiro, Brazil.

Bouassida, M., Guetif, Z., Buhan, P.E & Dormieux, L. (2003) *Estimation par une approche variationnelle du tassement d'une foundation sur sol renforcé par colonnes ballastées. Revue Française de Géotechnique*, 102, 21–22.

Bressani, L.A. (1983) *Contribution to the study of stress-strain-strength of Sarapuí clay (in Portuguese)*. Master's Thesis, PUC-Rio, Rio de Janeiro, Brazil.

Briançon, L., Delmas, P.H. & Villard, P. (2010) Study of the load transfer mechanisms in reinforced pile-supported embankments, Piled Embankments. In: IGS, *9th International Conference on Geosynthetics, Guarujá, Brazil*, vol. 4, pp. 1917–1924.

British Standards Institution (1995) BS 8006. *Code of practice for strengthened/reinforced soils and other fills*. London, UK, BSI.

British Standards Institution. (1999) BS 5930. *Code of practice for site investigations*. London, UK, BSI.

Brugger, P.J., Almeida, M.S.S., Sandroni, S.S., Brant, J.R., Lacerda, W.A. & Danziger, F.A.B. (1994) Geotechnical parameters of Sergipe Clay from Critical State Theory (in Portuguese). In: ABMS, *10th Brazilian Conference on Soil Mechanics and Foundation Engineering*, 1994, *Foz do Iguaçu, Brazil*, vol. 2, pp. 539–546.

Carlsson, B. (1987) *Reinforced soil – Principles for calculation*. Linköping, Terranova.

Carrilo, N. (1942) *Simple two and three dimensional cases in the theory of consolidation of soils. Journal of Math. and Phys.*, 21, 1–5.

Castro, J. & Karstunen, K. (2010) Numerical simulations of stone column installation. *Canadian Geotechnical Journal*, 47 (10), 1127–1138.

Castro, J. & Sagaseta, C. (2009) Field instrumentation of an embankment on stone columns. In: ISSMGE, *17th International Conference on Soil Mechanics and Geotechnical Engineering, Alexandria, Egypt*, CD-ROM.

Castro, J. (2008) *Análisis teórico de laconsolidación y deformaciónalrededor de columnas de grava*. Doctoral Thesis, Cantabria University, Santander, Spain.

Cedergren, H.R. (1967) *Seepage, drainage, and flow nets*. New York: John Wiley & Sons.

Chai, J. & Carter, J. (2011). Deformation analysis in soft ground improvement. *Geotechnical, geological and earthquake engineering*. Springer, vol. 18.

Chai, J, Miura, N. & Bergado, D.T., (2008). Preloading clayey deposit by vacuum pressure with cap-drain: analyses versus performance. *Geotextiles and Geomembranes*, 26 (3), 220–230.

Chai, J., Bergado, D.T. & Hino, T. (2010) FEM simulation of vacuum consolidation with CPVD for underconsolidated deposit. In: Almeida, M.S.S. (ed.) *Symposium New Techniques for Design and Construction in Soft Clays, 2010, Guarujá, Brazil*. São Paulo, Oficina de Textos. pp. 39–51.

Chandler, R.J. (1988) The in situ measurement of the undrained shear strength of clays using the field vane. In: Richards, A.F. (ed.). *Vane shear strength testing in soils: field and laboratory studies*, ASTM Publication, Philadelphia, United States, pp. 13–44.

Chen, R.H. & Chen, C.N. (1986) Permeability characteristics of prefabricated vertical drains. In: IGS, 3rd *International Conference on Geotextiles, Viena, Austria*, pp. 785–790.

Choobbasti, A., Zahmatkesh, A. & Noorzad, R. (2011) Performance of Stone Columns in Soft Clay: Numerical Evaluation. *Geotechnical and Geological Engineering*, 29 (5), 675–684.

Christopher, B., Holtz, R.D. & Berg, R.R. (2000) *Geosynthetic reinforced embankments on soft foundations, Soft Ground Technology*. In: J.L. Hanson e R.J. Termaat (eds.), ASCE – Special Publication, 112, 206–245.

Collet, H.B. (1978) *Vane tests on soft clays of Fluminense plains (in Portuguese)*. Master's. Thesis, COPPE/UFRJ, Rio de Janeiro, Brazil.

Collin, J.G. (2004) Column supported embankment design considerations. In: J.F. Labuz and J.G. Bentler (eds.), *52nd Annual Geotechnical Engineering Conference, Minnesota, USA, 2004*, pp. 51–78.

Costa Filho, L.M., Gerscovich, D.M., Bressani, L.A. & Thomaz, J.E. (1985) *Discussion on Journal of the Geotechnical Engineering Division*, ASCE, 111 (2), 259–262.

Costa Filho, L.M., Werneck, M.L.G. & Collet, H.B. (1977) The undrained strength of a very soft clay. In: ISSMGE, *9th International Conference on Soil Mechanics and Foundation Engineering*, Tokyo, Japan, vol. 1, 69–72.

Coutinho, R.Q. & Lacerda, W.A. (1987) Characterization – consolidation of Juturnaíba organic clay. In: *International Symposium on Geotechnical Engineering of Soft Soil*, Mexico, vol. 1. pp. 17–24.

Coutinho, R.Q. & Oliveira, A.T.J. (2000) Geotechnical properties of Recife soft clays. *Soils and Rocks*, ABMS, São Paulo, Brazil, 23 (3), 177–204.

Coutinho, R.Q. & Oliveira, J.T.R., França, A.E., Danziger, F.A.B. (1998) Piezocone tests on Ibura soft clay, Recife, Pernambuco State (in Portuguese). In: ABMS, *11th Brazilian Conference on Soil Mechanics and Foundation Engineering, 1998*, Brasília, Brazil, vol. 2, pp. 957–966.

Coutinho, R.Q. (1976) *Consolidation characteristics from radial drainage tests on a soft clay of Fluminense Lowlands (in Portuguese)*. Master's Thesis, COPPE/UFRJ, Rio de Janeiro, Brazil.

Coutinho, R.Q. (1976) *Radial drainage consolidation characteristics of Fluminense Plains soft clay (in Portuguese)*. Master's Thesis, COPPE/UFRJ, Rio de Janeiro, Brazil.

Coutinho, R.Q. (1986) *Monitored experimental embankment built until rupture over soft soils of Juturnaíba (in Portuguese)*. Doctoral Thesis, COPPE/UFRJ, Rio de Janeiro, Brazil.

Coutinho, R.Q. (2007) Characterization and Engineering properties of Recife soft clays – Brazil. In: Taylor & Francis (eds.), *Characterization and engineering properties of natural soils*. London, Taylor & Francis/Balkema, vol. 3, pp. 2049–2099.

Coutinho, R.Q. (2008) Field geotechnical Investigation and advances in practice. In: ABMS, *14th Brazilian Conference on Soil Mechanics and Foundation Engineering, Búzios, Rio de Janeiro, Brazil*, vol. 1. pp. 201–230.

Coutinho, R.Q., Almeida, M.S.S. & Borges, J.B. (1994) Analysis of the Juturnaíba. Embankment dam built on organic soft clays. In: ASCE, *Speciality Conference Settlement'94, Texas*, vol. 1.

Coutinho, R.Q., Oliveira, J.T.R. & Danziger, F.A.B. (1993) Geotechnical characterization of a soft clay from Recife (in Portuguese). Soils and Rocks, ABMS, São Paulo, Brazil, 16 (4), 255–266.

Coutinho, R.Q., Oliveira, J.T.R. & Oliveira, A.T.J. (1998) Quantification study of sampling quality of Brazilian soft clays – Recife and Rio de Janeiro (in Portuguese). In: ABMS, *10th Brazilian Conference on Soil Mechanics and Foundation Engineering, Brasília, Brazil*, vol. 2, 927–936.

Coutinho, R.Q., Oliveira, J.T.R. & Oliveira, A.T.J. (1998) Quantitative study of sample quality of brazilian soft clays – Recife and Rio de Janeiro (in Portuguese). In: ABMS, *11th Brazilian Conference on Soil Mechanics and Foundation Engineering, 1998, Brasília, Brazil*, vol. 2, pp. 927–936.

Crespo Neto, F.N. (2004) *Rate effect on strength from Vane test (in Portuguese)*. Master's Thesis, COPPE/UFRJ, Rio de Janeiro, Brazil.

Danziger, F.A.B. & Schnaid, F. (2000) Piezocone tests: procedures, recommendations and interpretation (in Portuguese). In: ABMS, *Brazilian Seminar on Field Investigation*, pp. 1–51.

Danziger, F.A.B. (1990) *Development of an equipment for piezocone tests: application to soft clays (in Portuguese)*. Doctoral Thesis, COPPE/UFRJ, Rio de Janeiro, Brazil.

Danziger, F.A.B., Almeida, M.S.S. & Sills, G.C. (1997) The significance of the strain path analysis in the interpretation of piezocone dissipation data. *Géotechnique*, UK, 47 (5), 901–914.

Davis, E.H. & Booker, J.R. (1973) *The effect of increasing strength with depth on the bearing capacity of clays*. Géotechnique, UK, 23, 551–563.

Departamento Nacional de Infraestruturas de Transportes (1998) DNER-PRO 381/98. *Design of embankments over soft soils for road works (in Portuguese)*, Brazil, DNIT.

Di Maggio, J.A. (1978) *Stone columns for highway construction*. U.S. Department of Transportation, Federal Highway Administration. Report number: FHWA-DP-46-1, 1978.

Di Prisco, C., Galli, A., Cantarelli, E. & Bongiorno, D. (2006) Georeinforced sand columns: small scale experimental tests and theoretical modeling. In: IGS, *8th International Conference on Geosynthetics*, Millpress, Rotterdam, Netherlands, pp. 1685–1688.

Dias, C.R.R. & Moraes, J.M. (1998) The experience about oft clays at the region of Laguna dos Patos Estuary and Port of Rio Grande. In: ABMS, *GEOSUL*, 1998, Porto Alegre, Brazil, pp. 179–96.

Dias, C.R.R. (2001) Geotechnical parameters and the influence of geological events – soft clays of Rio Grande State (in Portuguese). In: COPPE/UFRJ, *Workshop Properties of Brazilian Soft Soils, March 2001*, Rio de Janeiro, Brazil, CD-ROM.

Díaz-Rodríguez, J.A., Leroueil, S. & Alemán, J.D. (1992) Yielding of Mexico City clay and others natural clays. *Journal of the Geotechnical Engineering Division*, ASCE, 118 (7), 981–995.

Duncan, J.M. & Wright, S.G. (2005) *Soil strength and Slope Stability*. New York: John Wiley & Sons.

Dunnicliff, J. (1998) *Geotechnical instrumentation for monitoring field performance*. New York: John Wiley & Sons.

EBGEO (2010). Recommendations for Design and Analysis of Earth Structures using Geosynthetic Reinforcements, German Geotechnical Society (DGGT), Germany.

Egan, D., Scott, W. & McCabe, B.A. (2008) Observed installation effects of vibro-replacement stone columns in soft Clay. In: Karstunen, Leoni (eds.), *2nd International Workshop on Soft Soils – Focus on Ground Improvement, Glasgow, Scotland*, CRC Press, pp. 23–29.

Ehrlich, M. (1993) Design method for ballast of gravel over caps and piles (in Portuguese). *Soils and Rocks*, ABMS, São Paulo, Brazil, (16) 4, 229–234.

Ehrlich, M. & Becker, L.D.B. (2010) *Reinforced Soil Walls and Slopes: Design and Construction*, CRC Press, Balkema.

Ellstein, A. (1971) Settlement prediction through the sinking rate. *Revista Latino Americana de Geotecnia*, 1 (3), 231–237.

Empresa Brasileira De Pesquisa Agropecuária, Centro Nacional de Pesquisa de Solos. *Manual of methods for soil analysis (in Portuguese)*. 2nd ed., Rio de Janeiro, Brazil, Embrapa, 1997.

Escario, V. & Uriel, S. (1961) Determining the coefficient of consolidation and horizontal permeability by radial drainage. In: ISSMGE, *5th International Conference on Soil Mechanics and Foundation Engineering, Paris, France*, vol. 1, pp. 83–87.

Feijó, R.L. & Martins, I.M.S. (1993) Relationship between secondary compression, OCR and $K_0$(in Portuguese). In: COPPE/UFRJ, *COPPEGEO 93, Rio de Janeiro, Brazil*, pp. 27–40.

Feijó, R.L. (1991) *Relationship between secondary compression, overconsolidation ratio and earth pressure at rest* (in Portuguese). Master's Thesis, COPPE/UFRJ, Rio de Janeiro, Brazil.

Ferreira, C.A.M. (1991) *Analysis of piezometric data of an embankment on soft clay (in Portuguese)*. Master's Thesis, COPPE/UFRJ, Rio de Janeiro, Brazil.

Ferreira, S.R.M. & Coutinho, R.Q. (1988) Quantification of remolding effect on compressibility characteristics of the soft clay of Rio de Janeiro and Recife (in Portuguese). In: *Symposium on quaternary deposits of Brazilian coast plains: origin, geotechnical characteristics and field experience*, 3.55–3.69.

Filz, G.M. & Smith, M.E. (2006) *Design of bridging layers in geosynthetic-reinforced, column-supported embankments*. Virginia Transportation Research Council, Charlottesville. Report Number: VTRC 06-CR12.

Filz, G.M., Sloan, J., McGuire, M.P., Collin, J. & Smith, M.E. (2012) Column-Supported Embankments: Settlement and Load Transfer. In: ASCE, *Geotechnical Engineering State of Art and Practice, Keynote Lectures from Geo Congress 2012*, Reston, Virginia, 226, pp. 54–77

Francisco, G.M. (1997) *Seismic piezocone tests in soil (in Portuguese)*. Master's Thesis, PUC-Rio, Rio de Janeiro, Brazil.

Futai, M. M., Almeida, M. S. S & Lacerda, W. A. (2008) Laboratory behavior of Rio de Janeiro soft clays. Part 2: strength and yield. *Soils and Rocks*, 31 (2), 77–84.

Futai, M.M. (1999) *Theoretical and practical concepts on behaviour analysis of some Rio de Janeiro clays* (in Portuguese). Doctoral Seminar COPPE/UFRJ, Rio de Janeiro, Brazil.

Gäb, M., Schweiger, H.F., Kamrat-Pietraszewska, D. & Karstunen, M. (2009) Numerical analysis of a floating stone column foundation using different constitutive models. In: Karstunen, Leoni (eds.), *2nd International Workshop on Soft Soils – Focus on Ground Improvement, Glasgow, Scotland*, CRC Press, pp. 137–142.

Garcia, S.G.F. (1996) *Relationship between secondary consolidation and stress relaxation of a soft clay under oedometric compression* (in Portuguese). Master's Thesis, COPPE/UFRJ, Rio de Janeiro, Brazil.

Gebreselassie, B., Lüking, J. & Kempfert, H.G. (2010) Influence factors on the performance of geosynthetic reinforced and pile supported embankments, Piled Embankments. In: IGS, *9th International Conference on Geosynthetics, Guarujá, Brazil*, vol. 4, pp. 1935–1940.

Gerscovich, D.M. (1983) *Properties of the desiccated crust of the Sarapuí soft clay deposit* (in Portuguese). Master's Thesis, PUC-Rio, Rio de Janeiro, Brazil.

Gerscovich, D.M., Costa Filho, L.M. & Bressani, L.A. (1986) Proprieties of the dissecated crust of a soft clay deposit of the Fluminense Plains. In: ABMS, *8th Brazilian Conference on Soil Mechanics and Foundation Engineering*, Porto Alegre, vol. 2, 289–300.

Ghionna, V. & Jamiolkowski, M. (1981) Colonne di *ghiaia*. In: *Ciclo Di Conferenze Dedicate Ai Problemi Di Meccanica Dei Terreni E Ingegneria Delle Fondazioni Metodi Di Miglioramento Dei Terreni*, 10, Politecnico di Torino Ingegneria, Atti dell'Istituto di Scienza delle Costruzioni, n. 507.

Giroud, J.P. (1990) Functions and applications of geosynthetics in dams. In: *Water Power and Dam Construction*, vol. 42, n. 6, pp. 16–23.

Gneil, J. & Buazza, A. (2010) Construction of geogrid encased stone columns: a new proposal based on laboratory testing. *Geotextiles and Geomembranes*, 28 (1), 108–118.

Greenwood, D.A. (1970) Mechanical improvement of soils below ground surface. In: Institute of Civil Engineering, *Ground Engineering Conference*, London, pp. 9–20.

Han, J. & Ye, S.L. (2001) Simplified method for consolidation rate of stone column reinforced foundations. Journal of Geotechnical and Environmental Engineering, ASCE, 127 (7), 597–603.

Han, J. & Ye, S.L. (2002) A theoretical solution for consolidation rates of stone column-reinforced foundations accounting for smear and well resistance effects. *International Journal of Geomechanics*, ASCE, 2 (2), 135–151.

Han, J. (2010) Consolidation settlement of stone column-reinforced foundations in soft soils. In: Almeida, M.S.S. (ed.), *New techniques on soft soils, Guarujá, Brazil*. São Paulo, Oficina de Textos, pp. 167–177.

Hansbo, S. (1979) Consolidation of clay by bandshaped prefabricated vertical drains. *Ground Engineering*, 12 (5), 16–25.

Hansbo, S. (1981) Consolidation of fine-grained soils by pre-fabricated drains. In: ISSMGE, *10th International Conference on Soil Mechanics and Foundation Engineering*, Stockholm, vol. 3, pp. 677–682.

Hansbo, S. (1987) Facts and fiction in the field of vertical drainage. In: *Int. Symp. on Prediction and in Geot. Eng., Alberta, Canada*, pp. 61–72.

Hansbo, S. (2004) Band drains. In: Moseley, M.P., Kirsch, K. (eds.) *Ground improvement*, Taylor & Francis, pp. 4–56.

Head, K.H. (1982) *Manual of soil laboratory testing*. New York, John Wiley & Sons, vol. 2.

Hewlett, W.J. & Randolph, M.F. (1988) Analysis of piled embankment. Ground Engineering, 21 (3), 12–18.

Hight, D.W. (2001) Sampling effects in soft clay: an update on Ladd and Lambe (1963). In: ASCE, *Symposium on Soil Behaviour and Soft Ground Construction*. ASCE Geotechnical Special Publication, 119, pp. 86–121.

Hinchberger, S.D. & Rowe, R.K. (2003) Geosynthetic reinforced embankment on soft clay foundations: predicting reinforcement strains at failure. *Geotextiles and Geomembranes*, 21, 151–175.

Hird, C.C. & Moseley, V.J. (2000) Model study of seepage in smear zones around vertical drains in layered soil. *Geotechnique*, 50 (1), 89–97.

Holtz, R.D., Jamiolkowiski, M., Lancellota, R. & Pedroni, S. (1991) *Prefabricated Vertical Drains*. Londres, CIRIA, RPS 364.

Holtz, R.D., Shang, J.Q. & Bergado, D.T. (2001) Soil improvement. In: Kerry Rowe, R. (ed.). *Geotechnical and geoenvironment engineering handbook*. Norwel: Kluwer Academic Publishers.

Horgan, G.J. & Sarsby, R.W. (2002) The arching effect of soils over voids and piles incorporating geosynthetic reinforcement.In: IGS, *9th International Conference on Geosynthetics*, pp.373.

Houlsby, G.T. & Teh, C.I. (1988) Analysis of the piezocone in clay. In: *International Symposium on Penetration Testing*, ISOPT-1, Orlando, 2, Balkema Pub., Rotterdam, pp. 777–783.

Jamiolkowski, M. & Lancellotta, R. (1981) Consolidation by vertical drains-uncertainties involved in prediction of settlement rates. In: ISSMGE, Panel Discussion, *10th International Conference on Soil Mechanics and Foundation Engineering, Stockholm.*

Jamiolkowski, M., Lancellotta, R. & Wolski, W. (1983) Precompression and speeding up consolidation. In: *8th European Conference Soil Mechanics and Foundation Engineering, General Report, Special Session 6, Helsinki,* vol. 3, pp. 1201–1226.

Indraratna, B. & Redana, I.W. (1998) Laboratory determination of smear zone due to vertical drain installation. *Journal of Geotechnical Engineering, ASCE,* 125 (1), 96–99.

Indraratna, B., Sathananthan, I., Bamunawita, C. & Balasubramaniam, A.S. (2005) Theoretical and numerical perspectives and field observations for the design and performance evaluation of embankments constructed on soft marine clay. In: Indraratna, B., Chu, J. & Hudson, J.A. (eds.) *Elsevier Geo-Engineering Book Series, Ground Improvement – Case Histories,* Oxford, Elsevier, vol. 3, pp. 51–89.

Irex (2012) Recommandations pour la conception, le dimensionnement, l'exécution et le contrôle de l'amélioration: Projet national ASIRI (Recommendations for the design, construction and control of rigid inclusion ground improvements: ASIRI National Project), Presses des Ponts, Paris.

Jamiolkowski, M., Ladd, C.C., Germaine, J.T. & Lancellotta, R. (1985) New developments in field and laboratory testing of soils. In: ISSMGE, *11th International Conference on Soil Mechanics and Foundation Engineering,* San Francisco, USA, vol. 1, pp. 57–153.

Janbu, N. (1973) Slope stability computations. In: R.C. Hirschfeld, S.J. Poulos (ed.) *Embankment-dam engineering: Casagrande volume,* New York, John Wiley & Sons, pp. 47–86.

Jannuzzi, G.M.F. Characterization of the soft soil deposit of Sarapuí II using Field tests (in Portuguese). 2009. Master's Thesis – COPPE/UFRJ, Rio de Janeiro, 2009.

Jennings, K. & Naughton, P.J. (2010) Lateral deformation under the side slopes of piled embankments, Piled Embankments. In: IGS, *9th International Conference on Geosynthetics, 2010, Guarujá, Brazil,* vol. 4, pp. 1925–1933.

Jewell, R.A. (1982) A limit equilibrium design method for reinforced embankments on soft foundations. In: *International Conference on Geotextiles,* Las Vegas, USA, vol. 3, pp. 671–676.

Kabbaj, M., Tavenas, F. & Leroueil, S. (1988) In situ and laboratory stress-strain relationships. *Géotechnique,* 38 (1), 83–100.

Kavazanjian Jr., E. & Mitchell, J.K. (1984) Time dependence of lateral earth pressure. *Journal of Geotechnical Engineering,* New York, 110 (4), 530–533.

Kempfert, H.-G. & Gebreselassie, B. (2006) *Excavations and foundations in soft soils.* Berlin, Springer.

Kempfert, H.-G. (2003) Ground improvement methods with special emphasis on column-type techniques. In: Vermeer, Schweiger, Karstunen & Cudny (eds.), *International Workshop on Geotechnics of Soft Soils-Theory and Pratice, Noordwijkerhout, Netherlands,* pp. 101–112.

Kempfert, H.-G., Gobel, C., Alexiew & D., Heitz, C. (2004) German recommendations for reinforced embankments on pile-similar elements. In: EuroGeo3-*3rd European Geosynthetics Conference, Geotechnical Engineering with Geosynthetics,* pp. 279–284.

Kempfert, H.G., Jaup, A. & Raithel, M. (1997) Interactive behavior of a flexible reinforced sand column foundation in soft soils. In: ISSMGE, *14th International Conference on Soil Mechanics and Geotechnical Engineering,* Hamburg, Germany, pp. 1757–1760.

Kirsch, F. & Sondermann, W. (2001). Ground improvement and its numerical analysis. In: ISSMGE, *15th International Conference on Soil Mechanics and Geotechnical Engineering, Istanbul, Turkey,* Balkema, vol. 3, pp. 1775–1778.

Kitazume, M. (2005) The sand compaction pile method. Taylor & Francis.

Kjellman, W. (1952) Consolidation of clay soil by means of atmospheric pressure. In: Conference on Soil Stabilization, MIT, Cambridge, pp. 258–263.

Koerner, R.M. & Hsuan, Y.G. (2001) *Geosynthetics: characteristics and testing. Geotechnical and Geoenvironmental Engineering Handbook.* R.K. Rowe (ed.), pp. 173–196.

Koskinen, M., Karstunen, M., & Wheeler, S.J. (2002) Modelling destructuration and anisotropy of a natural soft clay. In: P. Mestat, (ed.), *5th European Conf. Numerical Methods in Geotechnical Engineering*, Presses de l'ENPC/LCPC, Paris, pp. 11–20.

Lacerda, W.A. & Almeida, M.S.S. (1995) Engineering properties of regional soils: residual soils and soft clays. In: *10th Panamerican Conference on Soil Mechanics and Foundation Engineering, State-of-the-art Lecture, Guadalajara, Mexico*, vol. 4, pp. 161–176.

Lacerda, W.A., Almeida, M.S.S., Santa Maria, P.E.L. & Coutinho, R.Q. (1995) Interpretation of radial consolidation tests. In: Yoshikuni, Kusabe (eds.), *International Symposium on Compression and Consolidation of Clayey Soils, Hiroshima, Japan.* Rotterdam, Balkema, vol. 2, pp.1091–1096.

Lacerda, W.A., Costa Filho, L.M., Coutinho, R.Q. & Duarte, A.R. (1977) Consolidation characteristics of Rio de Janeiro soft clay. In: *Conference on Geotechnical Aspects of Soft Clays, Bangkok*, pp. 231–244.

Ladd, C.C. (1991) Stability evaluation during staged construction. *Journal of Geotechnical Engineering*, ASCE, 117(4), 537–615.

Ladd, C.C. & De Groot, D.J. (2003) Recommended practice for soft ground site characterization: Casagrande Lecture. In: *12nd Panamerican Conference of Soil Mechanics*, Boston, USA.

Ladd, C.C. & Lambe, T.W. (1963) The strength of undisturbed clay determined from undrained tests. In: ASTM, *Symposium on Laboratory Shear Testing of Soils*, STP 361, pp. 342–371.

Law, K.T. (1985) Use o field vane tests under earth structures. In: ISSMGE, 11st *International Conference on Soil Mechanics & Foundation Engineering, San Francisco, USA*, vol. 2, pp. 893–898.

Leonards, G.A. & Girault, P. (1961) A study of the one-dimensional consolidation test. In: *5th International Conference on Soil Mechanics and Foundation Engineering*, Paris, vol. 1. pp. 213–218.

Leroueil, S. (1994) Compressibility of clays: fundamental and practical aspects. In: ASCE, *ASCE Specialty Conference on Settlement*, College Station, vol. 1, pp. 57–76.

Leroueil, S. (1997) Notes de cours: Comportement des massifs de sols. Université Laval, Québec, Canada.

Leroueil S. & Hight, D.W. (2003) Characterisation of soils for engineering. Characterisation and engineering properties of natural soils. In: T.S. Tan, K.K. Phoon, D.W. Hight & S. Leroueil (eds.) *International Workshop on Characterization and Engineering Properties of Natural Soils, Singapore*, vol. 1, pp. 255–36.

Leroueil, S. & Marques, M.E.S., (1996). Importance of strain rate and temperature effects in geotechnical engineering (state-of-the-art). In: ASCE, *Measuring and Modeling timedependent Soil Behaviour, Geotechnical Special Publication n° 61*, 1–60.

Leroueil, S. & Rowe, R.K. (2001) Embankments over soft soil and peat. In: Rowe, R.K. (ed.). *Geotechnical and Geoenvironmental Engineering Handbook.* USA: Kluwer Academic Publishers. pp. 463–499.

Leroueil, S. & Tavenas, F. (1986) Discussion on "Effective stress paths and yielding in soft clays below embankments", by D.J. Folkes & J. H. A. Crooks. *Canadian Geotechnical Journal*, 23 (3), 410–413.

Leroueil, S., Kabbaj, M., Tavenas, F. & Bouchard, R. (1985) Stress-strain – Strain rate relation for the compressibility of sensitive natural clays. *Géotechnique*, 35 (2), 159–180.

Leroueil, S., Magnan, J.P. & Tavenas, F. (1985) *Remblais sur argiles molles.* Paris, Technique et Documentation Lavoisier.

Leroueil, S., Tavenas, F. & Le Bihan, J.P. (1983) Propriétés caractéristiques des argiles de L'est du Canada. *Canadian Geotechnical Journal*, 20, 681–705.

Leroueil, S., Tavenas, F., Mieussens, C. & Peignaud, M. (1978) Construction pore pressures in clay foundations under embankments. Part II: generalized behaviour. *Canadian Geotechnical Journal*, 15 (1), 66–82.

Leroueil, S. & Watabe, Y. (2012). Consolidation of clays and ground improvement. In: Buddhima Indraratna, Cholachat Rujikiatkamjorn & Jayan Vinod (eds.) *International conference on ground improvement and ground control: transport infrastructure development and natural hazards mitigation, 2012, Wollongong, Australia*, vol. 1.

Leshchinsky, D., Leshchinsky, O., Ling, I. & Gilbert, P. (1996) Geosynthetic tubes for confining pressurized slurry: some design aspects. Journal of Geotechnical Engineering, 682–690.

Lima, B.T. (2012) *Study of the use of stone columns on very soft clayey soils (in Portuguese)*. Doctoral Thesis – COPPE-UFRJ, Rio de Janeiro, Brazil.

Lima, B.T. & Almeida, M.S.S. (2009) Light fill with EPS over a very soft soil at Rio de Janeiro (in Portuguese). In: *3rd Conferencia Sudamericana de Ingenieros Geotécnicos Jóvenes, Desafíos y avances de la Geotecnia Joven en Sudamérica, Córdoba, Argentina*, vol. 1. pp. 153–156.

Liu, H.L. & Chu, J. (2009) A new type of prefabricated vertical drain, with improved properties. *Geotextiles and Geomembranes*, 27 (2), pp. 152–155.

Long, M. & Phoon, K.K. (2004) General report: innovative technologies and equipment. Geotechnical and geophysical site characterization. In : A. Viana da Fonseca & P.W. Mayne (eds.) *2nd International Conference on Site Characterization, Porto, Portugal*, vol. 1, pp. 625–635.

Long, R.P. & Carey, P.J. (1978) Analysis of settlement data from sand drained areas. *Transportation Research Record, Washington*, n. 678, pp. 37–40.

Low, B.K., Wong, K.S. & Lim, C. (1990) Slip circle analysis of reinforced embankment on soft ground. *Journal of Geotextiles and Geomembranes*, 9 (2), 165–181.

Low, B.K., Tang, S.K. & Choa, V. (1994) Arching in piled embankements. *Journal of Geotechnical and Geoenvironmental Engineering*, ASCE, 120 (11).

Lunne, T., Berre, T. & Strandvik, S. (1997) Sample disturbance effects in softs low plastic Norwegian clay. In: Almeida (ed.), *Recent developments in soil mechanics*, Rotterdam: Balkema, pp. 81–102.

Lunne, T., Robertson, P.K. & Powell, J.J.M. (1997) *Cone penetration testing in geotechnical practice*. London, Spon Press.

Macedo, E. O. (2004) *Research on in situ undrained strength from T-bar tests (in Portuguese)*. Master's Thesis – COPPE/UFRJ, Rio de Janeiro, Brazil.

Magnan, J.P. (1983) Théorie et pratique des drains verticaux, Technique et Documentation. Paris, Lavoisier.

Magnani H.O., Almeida, M.S.S. & Ehrlich, M. (2009) Behaviour of two reinforced test embankments on soft clay taken to failure. *Geosynthetics International*, 3, 127–138.

Magnani, H.O. (2006) *Behaviour of reinforced embankments over soft soils built until rupture (in Portuguese)*. Doctoral Thesis, COPPE-UFRJ, Rio de Janeiro, Brazil.

Magnani, H.O., Almeida, M.S.S. & Ehrlich, M. (2008) Construction stability evaluation of reinforced embankments over soft soils. In: IGS, *1st Pan American Geosynthetics Conference & Exhibition*, Cancun. CD-ROM, pp. 1372–1381.

Magnani, H.O., Ehrlich, M. & Almeida, M.S.S. (2010) Embankments over soft slay deposits: contribution of basal reinforcement and surface sand layer to stability. *Journal of Geotechnical and Geoenvironmental Engineering*, 136, 260–264.

Mandel, J. & Salençon, J. (1972) Force Portance d'ún Sol sur une Assise Rigide: Étude Théorique. *Géotechnique*, 22, 79–93.

Marques, M.E.S. (2001) *Influence of viscosity on the laboratory and field behaviour of clayey soils (in Portuguese)*. Doctoral Thesis, COPPE/UFRJ, Rio de Janeiro, Brazil.

Marques, M.E.S. & Lacerda, W.A. (2004) Geotechnical characterization of a fluvial-marine deposit at Navegantes, Santa Catarina State (in Portuguese). In: *4th Seminar of the geotechnical engineering practice in South Brazil, Curitiba*, pp. 31–38.

Marques, M.E.S. & Leroueil, S. (2005) Preconsolidating clay deposit by vacuum and heating in cold environment. In: Indraratna, B., Chu, J. & Hudson, J.A. (eds.) *Ground Improvement – Case Histories*. Elsevier Geo-Engineering Book Series, vol. 3, pp. 1045–1063.

Marques, M.E.S., Oliveira, J.R.M.S. & Souza, A.I. (2008) Characterization of a sedimentar soft deposit at Porto Velho (in Portuguese). In: ABMS, *14th Brazilian Conference on Soil Mechanics and Foundation Engineering, 2008, Búzios, Rio de Janeiro, Brazil*.

Marques, M.E.S., Lima, B.T., Oliveira, J.R.M.S., Antoniutti Neto, L. & Almeida, M.S.S. (2008) Geotechnical characterization of a compressible deposit at Itaguaí, Rio de Janeiro State (in Portuguese). In: *4th Congresso Luso-brasileiro de Geotecnia, Coimbra, Portugal*.

Martins, I.S.M. & Lacerda, W.A. (1985) A theory for consolidation with secondary compression. In: ISSMGE, *11th International Conference on Soil Mechanics and Foundation Engineering, San Francisco, USA*, pp. 567–570.

Martins, I.S.M. (2005) *Some considerations about secondary consolidation (in Portuguese)*. Lecture at Clube de Engenharia, Rio de Janeiro, Brazil.

Martins, I.S.M., Santa Maria, P.E.L. & Lacerda, W.A. (1997) A brief review about the most significant results of COPPE research on rheological behaviour of saturated clays subjected to one- dimensional strain. In: Almeida, M.S.S. (ed.). *Recent developments in soil mechanics*. Rotterdam, Balkema, pp. 255–264.

Mason, J. (1982) *Obras portuárias*. 2nd ed, Rio de Janeiro, Brazil, Campus.

Massad, F. (2003) *Obras de terra: curso básico de geotecnia*. São Paulo, Oficina de Textos.

Massad, F. (2009) *Solos marinhos da Baixada Santista – Características e propriedades técnicas*. São Paulo: Oficina de Textos.

Massad, F. (1999) Santos Lowlands: implications of the geological history on foundation design (in Portuguese). *Soils and Rocks*, Brazil 22(1), 3–49.

Mayne, P.W. & Mitchell J.K. (1988) Profiling of overconsolidation ratio in clays by field vane. *Canadian Geotechnical Journal*, 25, 150–157.

Mc Guire, Filz, G.M. & Almeida, M.S.S. (2009) Load-Displacement Compatibility Analysis of a Low-Height Column-Supported Embankment. In: *Contemporary Topics in Ground Modification, Problem Soils, and Geo-Support, GSP, 187, ASCE, Orlando, FL*, pp. 225–232.

McCabe, B.A., Mcneill, J.A. & Black, J.A. (2007) Ground improvement using the vibro-stone column technique. *Joint meeting of Engineers Ireland West Region and the Geotechnical Society of Ireland*.

McCabe, B.A., Nimmons, G.J. & Egan, D. (2009) A review of field performance of stone columns in soft soils. *Journal of Geotechnical Engineering – Institution of Civil Engineers*, ICE,UK, 162 (6), 323–334.

Mcguire, M.P. & Filz, G.M. (2008) Quantitative comparison of theories for geosynthetic reinforcement of column-supported embankments. In: IGS, *1st Pan American Geosynthetics Conference & Exhibition, 2–5 March 2008, Cancun, México*.

Mcguire, M.P., Sloan, J., Collin, J. & Filz, G.M. (2012) Critical Height of Column-Supported Embankments from Bench-Scale and Field-Scale Tests In: ISSMGE–TC 211, *International Symposium on Ground Improvement IS-GI, Brussels, 31 May & 1 June 2012*.

McKenna, J.M., Eyre, W.A. & Wolstenholme, D.R. (1975). Performance of an embankment supported by stone columns in soft ground. *Geotechnique*, 25 (1), 51–59.

Mestat, P., Magnan, J.P. & Ddhouib, A. (2006) Results of the settlement prediction exercise of an embankment founded on soil improved by stone columns. In: Schweiger (ed.) *Numerical Methods in Geotechnical Engineering—NUMGE 06, Graz, Austria*. Taylor & Francis Group, London, pp. 471–476.

Mello, L.G., Mandolfo, M., Montez, F., Tsukahara, C.N. & Bilfinger, W. (2008) First use of geosynthetic encased sand columns in South America. In: *1st Pan American Geosynthetics Conference & Exhibition, Cancún, Mexico*. CD-ROM.

Mello, L.G., Schnaid, F. & Gaspari, G. (2002) Geotechnical characteristics of Natal coastal clays and its influence on the works of expansion of the port (in Portuguese). *Soils e Rocks, São Paulo*, 5 (1), 59–71.

Mello, M.A. (2013) *Deep radial consolidation aplied on compressible soil of Lagoa Rodrigo de Freitas, Rio de Janeiro (in Portuguese)*. Master's Thesis, Military Institute of Engineering, Rio de Janeiro, Brazil.

Mesri, G. (1975) Discussion on "new design procedure for stability of soft clays". *Journal of Geotechnical Engineering*, ASCE, 101(4), 409–412.

Mitchell, J.K. (1964) Shearing resistance of soils as a rate process. *Journal of the Soil Mechanics and Foundation Division*, ASCE, 90 (1), 29–61.

Mitchell, J.K. (1993). *Fundamentals of Soil Behaviour*, John Wiley, Sons, New York.

Mitchell, J.K. & Huber, T.R. (1985) Performance of a Stone Column Foundation. *Journal of Geotechnical Engineering*, ASCE, 111, 2.

Moormann, C. & Jud, H. (2010) Foundation of a coal/coke stockyard on soft soil with geotextile encased columns and horizontal reinforcement. In: IGS, *9th International Conference on Geosynthetics, Guarujá, Brazil*, pp. 1905–1908.

Murugesan, S. & Rajagopal, K. (2006) Geosynthetic-encased stone column: numerical evaluation. *Geotextiles and Geomembranes*, 24 (6), 349–358.

Murugesan, S. & Rajagopal, K. (2010) Studies on the Behavior of Single and Group of Geosynthetic Encased Stone Columns, *Journal of Geotechnical and Geoenvironmental Engineering*, ASCE, 136 (10), 129–139.

Nagaraj, T.S. & Jayadeva, M.S. (1983) Critical reappraisal of plasticity index of soil. *Journal of the Geotechnical Engineering Division*, ASCE, vol. 109 (7), 994–1000.

Nascimento, C.M.C. (2009) *Assessment of alternatives for executive processes to build embankments of urban roads on soft soils (in Portuguese)*. Master's Thesis, Military Institute of Engineering, Rio de Janeiro, Brazil.

Nascimento, I.N.S. (1998) *Development and utilization of in situ electrical vane equipment (in Portuguese)*. Master's Thesis, COPPE/UFRJ, Rio de Janeiro, Brazil.

Oliveira, A.T.J. & Coutinho, R.Q. (2000) Utilization of *in situ* electrical vane equipment on a soft clay of Recife (in Portuguese). In: *Brazilian Seminar of Field Investigation, São Paulo, Brazil*.

Oliveira, A.T.J. (2000) *Utilization of in situ electrical vane equipment on soft clays of Recife (in Portuguese)*. Master's Thesis, Federal University of Pernambuco, Recife, Brazil.

Oliveira, J.T.R. (1991) *Piezocone tests on a soft clay deposit of Recife clay (in Portuguese)*. Master's Thesis, COPPE/UFRJ, Rio de Janeiro, Brazil.

Oliveira, J.T.R. (2006) Geotechnical parameters of a soft clay of Suape Port, Pernambuco State (in Portuguese). In: ABMS *12ndBrazilian Conference on Soil Mechanics and Foundation Engineering, Curitiba*, vol 1, pp. 13–18.

Onoue, A., Ting, N.H., Germaine, J.T. & Whitman, R.V. (1991) Permeability of disturbed zone around vertical drains In: ASCE, *Geotechnical Congress, Colorado, USA*, pp. 879–890.

Orleach, P. (1983) *Techniques to evaluate the field performance of vertical drains*. Master's Thesis, MIT, Cambridge, Mass., USA.

Ortigão J.A.R. & Almeida M.S.S. (1988) Geotechnical Ocean Engineering. In:*Civil Engineering Practice*. Technomic Publishing CO., INC. vol. 3, pp. 267–331.

Ortigão, J. A. R., Werneck, M. L. G. & Lacerda, W. A. (1983) Embankment failure on clay near Rio de Janeiro. *Journal of the Geotechnical Engineering Division*, ASCE, 109 (11), 1460–1479.

Ortigão, J.A.R. & Collet, H. B. (1986) The elimination of friction errors on vane tests (in Portuguese). *Soils and Rocks*, ABMS, 9 (2), 33–45.

Ortigão, J.A.R. (1980). *Failure of a trial embankment on Rio de Janeiro gray clay (in Portuguese)*. Doctoral Thesis, COPPE/UFRJ, Rio de Janeiro, Brazil.

Pacheco Silva, F. (1953) Shearing strength of a soft clay deposit near Rio de Janeiro. *Géotechnique*, 3, 300–306.

Pacheco Silva, F. (1970) A new graphic construction for determination of preconsolidation pressure of a soil sample (in Portuguese). In: ABMS, *4th Brazilian Conference on Soil Mechanics and Foundation Engineering,*, Rio de Janeiro, Brazil, vol. 2 (I) , pp. 225–232.

Palmeira, E.M. & Almeida, M.S.S. (1979) Actualization of the program BISPO for stability analysis of slopes (in Portuguese). Report, IPR/DNER.

Palmeira, E.M. & Ortigão, J.A.R. (1981) Construction and performance of a full profile settlement gauge: profilometer for embankments. *Soils and Rocks*, 4 (2).

Palmeira, E.M. & Ortigão, J.A.R. (2004) Aplications of reinforcement – embankmnts over soft soils (in Portuguese). In: Vertematti, J.C. (ed.). *Brazilian Manual of Geosynthetics*. São Paulo, Edgar Blücher .

Parry, R.H.G. & Wroth, C.P. (1981) Shear properties of clay, in soft clay Engineering. In: E.W. Brand & R.P. Brenner (eds), Elsevier, Amsterdam, pp. 311–64.

Parry, R.H.G. (1972) Stability analysis of low embankments on soft clays. In: *Roscoe Memorial Symposium*, Cambridge University, pp. 643–668.

Pilarczyk, K.W. (2000) Geosynthetics and geosystems in hydraulic and coastal engineering. Rotterdam, Balkema.

Pilot, G. & Moreau, M. (1973) La stabilité des remblais sur sols mous. Abaques de Calcul, Eyrolles, Paris.

Pinto, C.S. (1966) Capacidade de carga de argila com coesão crescente com a profundidade. *Jornal de Solos*, 3(1) 21–44.

Pinto, C.S. (1994) Embankments on Lowlands (in Portuguese). In: Negro Jr., A., Ferreira, A. A., Alonso, U. R., Luz, P. A. C., Falconi, F. F. & Frota, R. Q. (eds.). *Solos do litoral do Estado de São Paulo*. 1st ed., São Paulo, ABMS, pp. 235–264.

Pinto, C.S. (2000) *Curso básico de mecânica dos solos*. São Paulo: Oficina de Textos.

Pinto, C.S. (2001). Considerations of Asaoka's method (in Portuguese). *Soils and Rocks*, ABMS, ABGE, 24 (1), 95–100.

Pinto, C.S., (1992) Topics of the contribution of Pacheco Silva and considerations of undrained strength of clays (in Portuguese). 1st Pacheco Silva Conference. *Soils and Rocks*, ABMS, 15 (2), 49–87.

Potts, V. & Zdravkovic, L. (2010) Finite-element study of arching behaviour in reinforced fills. In: Proceedings of the Institution of Civil Engineers – ICG. Ground Improvement 163. November, 2010.

Poulos, H.G. & Davis, E.H. (1974) Elastic solutions for soil and rock mechanics. New York: *John Wiley & Sons*.

Priebe, H.J. (1978) Abschatzung des Scherwderstandes eines durch Stopverdichtung verbesserten Baugundes. *Die Bautechnik*, 15 (8), 281–284.

Priebe, H.J. (1995) The design of vibro replacement. *Ground Engineering*, 31–37.

Pulko, B. & Majes, B. (2005) Simple and accurate prediction of settlements of stone column reinforced soil In: *16th International Conference on Soil Mechanics and Geotechnical Engineering*, Osaka, Japan, vol. 3, pp.1401–1404.

Raithel, M. & Kempfert, H.-G. (2000) Calculation models for dam foundations with geotextile coated sand columns. In: *GEOENG*, 2000, Melbourne, p. 347.

Raithel, M. (1999) Zum Trag- und Verformungsverhalten von geokunststoffummantelten Sandsäulen In: *Series Geotechnics*, University of Kassel, n. 6.

Raithel, M., Kirchner, A., Schade, C., & Leusink, E. (2005) Foundation of constructions very soft soils with geotextile encased columns – state of the art. *Geotechnical Special Publication*, 130–142, Geo-Frontiers 1867–1877.

Raju, V. R., Wegner, R. & Godenzie, D. (1998) Ground Improvement using vibro techniques – case histories from S. E. Asia. In: *Ground Improvement Conference, Singapore*, Keller Publications. CD-ROM.

Raju, V.R. & Sondermann, W. (2005) Ground improvement using deep vibro techniques. In: Indraratna, B., Chu, J., Hudson, J. A. (eds.). *Elsevier Geo-Engineering Book Series, v. 3, Ground Improvement – Case histories*. Oxford, Elsevier, pp. 601–638.

Ramos, O.G. & Niyama, S. (1994) Port works (in Portuguese). In: Negro Jr., A., Ferreira, A. A., Alonso, U. R., Luz, P. A. C., Falconi, F. F. & Frota, R. Q. (eds.). *Solos do litoral do Estado de São Paulo*. 1st ed., São Paulo, ABMS, pp. 265–288.

Randolph, M. F. (2004) Characterization of soft sediments for offshore applications geotechnical and geophysical site characterization. In : A. Viana da Fonseca & P.W. Mayne (eds.) *2nd International Conference on Site Characterization, Porto, Portugal*, vol. 1, pp. 209–232.

Rathgeb, E. & Kutzner, C. (1975) Some applications of the vibro-replacement process. *Géotechnique*, 31 (1), 43–157.

Remy, J.P.P, Martins, I.S.M., Santa Maria, P.E.L., Aguiar, V.N. & Andrade, M.E.S. (2010) The Embraport pilot embankment – primary and secondary consolidations of Santos soft clay with and without wick drains – Part 2. In: Almeida, M.S.S. (ed.) *New Techniques on Soft Soils. São Paulo*, Oficina de Textos, pp. 311–330.

Riccio, M.V.F. (2007) *Behavior of a reinforced soil built with fine grain size tropical soil*. Doctoral Thesis, COPPE/UFRJ, Rio de Janeiro, Brazil.

Rixner, J.J., Kreaemer, S.R. & Smith, A.D. (1986) *Prefabricated vertical drains. vol. 1*. U.S. Department of Transportation, Federal Highway Administration, Washington, DC, USA, Report number: FHWA-RD-86/168.

Robertson, P.K. (1990) Soil classification using the cone penetration test. *Canadian Geotechnical Journal*, 27 (1), 151–158.

Robertson, P.K., Sully, J.P., Woeller D.J., Lunne, T., Powell, J.J.M. & Gillespie, D.G. (1992) Estimating coefficient of consolidation from piezocone tests. *Canadian Geotechnical Journal*, 29 (4) 539–550.

Rocha Filho, P. & Alencar, J.A. (1985) Piezocone tests in the Rio de Janeiro soft clay deposit. In: ISSMGE, *11th International Conference on Soil Mechanics and Foundation Engineering*, San Francisco, USA, vol. 2, pp. 859–862.

Roscoe, K.H. & Burland, J.B. (1968). On the generalized stress – strain behavior of wet clay. In: J. Heyman & F. Leckie (ed.), *Engineering Plasticity*, Cambridge University Press, Cambridge, pp. 535–609.

Roscoe, K.H., Schofield, A.N. & Wroth, C.P. (1958). On the yielding of soils. *Géotechnique*, 8 (1), 22–53.

Rowe, R.K. & Soderman, K.L. (1985) An approximate method for estimating the stability of geotextile embankments. *Canadian Geotechnical Journal*, 22 (3), 392–398.

Russell, D. & Pierpoint, N. (1997) An assessment of design methods for piled embankments. *Ground Engineering*, 30 (11), 39–44.

Samara, V., Barros, J.M.C., Marco, L.A.A., Belincanta, A. & Wolle, C.M. (1982) Some properties of marine clays of Santos Lowlands (in Portuguese). In: ABMS, *8th Brazilian Conference on Soil Mechanics and Foundation Engineering, Recife, Brazil*, vol. 4, pp. 301–318.

Sandroni, S. S. (1993) About the use of vane tests in embankment design (in Portuguese). Soils and Rocks, 16 (3), 207–213.

Sandroni, S.S. (2006) About the Brazilian practice of geotechnical design of road embankments over very soft soils (in Portuguese). In: ABMS, *13th Brazilian Conference on Soil Mechanics and Foundation Engineering, 2006, Curitiba, Brazil,* CD-ROM.

Sandroni, S.S. & Bedeschi, M.V.R. (2008) Aterro instrumentado da área C – Uso de drenos verticais em depósito muito mole da Barra da Tijuca, Rio de Janeiro. In: ABMS, *14th Brazilian Conference on Soil Mechanics and Foundation Engineering, Búzios, Rio de Janeiro, Brazil,* CD-ROM.

Sandroni, S.S. & Deotti, L.O.G. Instrumented test embankments on piles and geogrid platforms at the Panamerican Village, Rio de Janeiro. In: *Pan American Geosynthetics Conference & Exhibition, 1, 2008, Cancún, Mexico. Proceedings.* Cancún, 2008. 1 CDROM.

Sandroni, S.S., Lacerda, W.A. & Brandt, J. R. (2004) Método dos volumes para controle de campo da estabilidade de aterros sobre argilas moles. *Soils and Rocks,* 27 (1), 25–35.

Sandroni, S.S., Brugger, P.J, Almeida, M.S.S. & Lacerda, W.A. (1997) Geotechnical properties of Sergipe clay. In: *International Symposium on Recent Developments in Soil and Pavement Mechanics, Rio de Janeiro, Brazil,* pp. 271–277.

Saye, R. (2001) Assessment of soil disturbance by the installation of displacement sand drains and prefabricated vertical drains. In: ASCE, *Geotechnical Special Publication,* n. 119, pp. 325–362.

Schmertmann, J.H. (1955) The undisturbed consolidation behaviour of clay. In: ASCE, *Transactions ASCE,* vol. 120, pp. 1201–1227.

Schmidt, C.A.B. (1992) *A settlement analysis by Asaoka's method with a probabilistic approach* (in Portuguese). Master's Thesis, COPPE/UFRJ, Rio de Janeiro, Brazil.

Schnaid, F. (2000) *Ensaios de campo e suas aplicações à engenharia de fundações.* São Paulo, Oficina de Textos.

Schnaid, F. (2008) Investigação geotécnica em maciços naturais não-convencionais. In: 4° *Congresso Luso-Brasileiro de Geotecnia, Coimbra,* Portugal, pp. 17–40.

Schnaid, F. (2009) *In situ testing in geomechanics.* 1st ed. Oxon: Taylor & Francis.

Schnaid, F., Nacci, D & Militittsky, J. (2001) *Aeroporto Salgado Filho – Infraestrutura civil e geotécnica.* 1st ed. Porto Alegre, Sagras.

Schnaid, F., Sills, G.C., Soares, J.M.D. & Byirendam, Z. (1997) Predictions of the coefficient of consolidation from piezocone tests. *Canadian Geotechnical Journal,* 34 (2), 143–159.

Schober, W. & Teindel, H. (1979) Filter criteria for geotextiles. Design parameters in geotechnical engineering. In: BGS, London, U.K., vol. 7, pp. 168–178.

Schofield, A. & Wroth, C.P. (1968). *Critical State Soil Mechanics.* London, McGraw-Hill.

Scott, R.F. (1961) New Method of Consolidation Coefficient Evaluation. *Journal of the Soil Mechanics and Foundation Division,* ASCE, 87 (1), 29–39.

Sills, G.C., Almeida, M.S.S. & Danziger, F.A.B. (1988) Coefficient of consolidation from piezocone dissipation tests in very soft clay. In: *International Symposium on Penetration Tests, Orlando, Florida,* vol. 2, pp. 967–974.

Six, V., Mroueh, H., Shahrour, I. & Bouassida, M., (2012) Numerical Analysis of Elastoplastic Behavior of Stone Column Foundation. *Geotechnical and Geological Engineering,* 1–13.

Smith, A. & Rollins, K. (2012) Minimum effective PV drain spacing from embankment field tests in soft clay. In: ISSMGE, *17th International Conference on Soil Mechanics and Geotechnical Engineering, Alexandria, Egypt,* pp. 2184–2187.

Soares, J.M.D., Schnaid, F. & Bica, A.V.D. (1997) Determination of the characteristics of a soft clay deposit in southern Brazil. In: *International Symposium on Recent Developments in Soil and Pavement Mechanics, Rio de Janeiro, Brazil,* pp. 297–302.

Soares, J.M.D., Schnaid, F. & Bica, A.V.D. (1994) Strength properties of a clay deposit from field tests. In: ABMS, *10th Brazilian Conference on Soil Mechanics and Foundation Engineering, Foz do Iguaçu, Brazil,* vol. 2, pp. 573–580.

Soares, M.M., Almeida, M.S.S. & Danziger, F.A.B. (1987) Piezocone research at COPPE/UFRJ. In: 6th *International Symposium on Offshore Engineering, Rio de Janeiro*, pp. 226–242.

Soares, M.M., Lunne, T., Almeida, M.S.S. & Danziger, F.A.B. (1986) Dilatometer test on soft soil (in Portuguese). In: ABMS, *8th Brazilian Conference on Soil Mechanics and Foundation Engineering, Porto Alegre, Brazil*, vol. 2, pp. 89–98.

Spotti, A.P. (2006) *Monitored reinforced embankment over piles over soft soil (in Portuguese)*. Doctoral Thesis, COPPE/UFRJ, Rio de Janeiro.

Springman, S.M., Laue, J., Askarinejad, A. & Gautray, J.N.F. (2012) On the design of ground improvement for embankements on soft ground. In: Buddhima Indraratna, Cholachat RujiKiatkamjorn & Jayan Vinod (eds.) *International conference on ground improvement and ground control: transport infrastructure development and natural hazards mitigation,*, Wollongong, Australia, vol. 1, pp. 67–83.

Stewart, D. P. & Randolph, M. F. (1991) A new site investigation tool for the centrifuge. H. Y. Ko, (Ed.). In: *Int. Conf. Centrifuge*, Rotterdam, A. A. Balkema, pp. 531–538.

Tan, S.A., Tjahyono, S. & Oo, K.K. (2008) Simplified plane-strain modeling of stone-column rein- forced ground. Journal of Geotechnical and Geoenvironmental Engineering, 134 (2), 185–194.

Tan, S.B. (1971) Empirical method for estimating secondary and total settlements. In: *Asian Regional Conf. on Soil Mech. and Foundation Engineering, Bangkok*, vol. 2. pp. 147–151.

Tavenas, F. & Leroueil, S. (1987) Laboratory and in situ stress-strain-time behaviour of soft clays: a state- of-the-art. In: *Simposio Internacional de Ingenería Geotécnica de Suelos Blandos, Mexico City, Mexico*, pp. 1–41.

Tavenas, F., Mieussens, C. & Bourges, F. (1979) Lateral displacements in clay foundations under embankments. *Canadian Geotechnical Journal*, 16 (3), 532–550.

Taylor, D.W. & Merchant, W. (1940) A theory of clay consolidation accounting for secondary compression. *Journal of Mathematics and Physics*, 19 (3), 167–185.

Terzaghi, K. & Frohlich, O.K. (1936) *Theorie der Setzung von Tonschichten*. Viena: Franz Deuticke.

Terzaghi, K. (1943) *Theoretical soil mechanics*. New York, John Wiley & Sons.

Thornburn, S. (1975) Building structures supported by stabilized ground. *Géotechnique*, 25 (1), 83–94.

Tschebotarioff, G.P. (1973) *Foundations, retaining and earth structures. The art of design and construction and its scientific basis in soil mechanics*. Tokyo, McGraw-Hill Kogakusha.

Van Der Stoel, A.E.C., Brok, C., De Lange, A.P. &Van Duijnen, P.G. (2010) Construction of the first railroad widening in the Netherlands on a Load Transfer Platform (LTP), Piled Embank- ments. In: IGS, *9th International Conference on Geosynthetics, Guarujá, Brazil*, vol. 4, pp. 1969–1972.

Van Dorp, T. (1996) Building on EPS geofoam in the "Low-Lands" experiences in the Nether- lands. In: *International Symposium on EPS Construction Method*, Tokyo, Japan, pp. 59–69.

Van Eekelen, S.J.M., Bezuijen, A. & Alexiew, D. (2010) The Kyoto road piled embankment: $3^{1}/_{2}$ years of measurements, Piled Embankments, In: IGS, *9th International Conference on Geosynthetics, Guarujá, Brazil*, vol. 4, pp. 1941–1944.

Van Eekelen, S.J.M., Jansen, H.L., Van Duijnen, P.G., De Kant, M., Van Dalen, J.H., Brugman, M. H.A., Van Der Stoel, A.E.C. & Peters, M.G.J.M. (2010)The Dutch design guideline for piled embankments, Piled Embankments. In: IGS, *9th International Conference on Geosynthetics, Guarujá, Brazil*, vol. 4, pp. 1911–1916.

Van Impe, W. & Silence, P. (1986). Improving of the bearing capacity of weak hydraulic fills by means of geotextiles. In: IGS, *Proceedings of the 3rd International Conference on Geotextiles, Vienna, Austria*, pp. 1411–1416.

Varaksin, S. (2010) Vacuum consolidation, vertical drains for the environment friendly consol- idation of very soft polluted mud at the Airbus A-380 factory site. In: Almeida, M.S.S. (ed.)

*Symposium New Techniques for Design and Construction in Soft Clays, Guarujá, Brazil.* São Paulo, Oficina de Textos, pp. 87–102.

Vieira, M.V.C.M (1994) *Dilatometer tests in soft Sarapuí clay (in Portuguese).* Master's Thesis, COPPE/UFRJ, Rio de Janeiro, Brazil.

Wang, G. (2009) Consolidation of soft clay foundations reinforced by stone columns under time-dependent loadings. *Journal of Geotechnical and Environmental Engineering,* ASCE, 135 (12), 192–193.

Weber, T.M., Springman, S.M., Gäb, M., Racansky, V. & Schweiger, H.F. (2009). Numerical modelling of stone columns in soft clay under an embankment. In: Karstunen, Leoni (eds.), *2nd International Workshop on Soft Soils – Focus on Ground Improvement, Glasgow, Scotland,* CRC Press, pp. 305–311.

Wheeler, S.J., Näätänen, A., Karstunen, M. & Lojander, M. (2003) An anisotropic elasto-plastic model for soft clays. *Canadian Geotechnical Journal,* 40 (2), 403–418.

Wissa, E.Z., Christian, J.T., Davis, E.H. & Heiberg, S. (1971) Consolidation at constant rate of strain. *Journal of the Soil Mechanics and Foundation Division,* ASCE, 97 (10), 1393–1413.

Wood, D.M. (1990) *Soil Behavior and Critical State Soil Mechanics.* Cambridge, Cambridge University Press.

Wroth, C.P. (1984) The interpretation of in situ soil tests. *Géotechnique,* 34 (4), 449–489.

Xiao, D. (2000) *Consolidation of soft clay using vertical drains.* Doctoral Thesis, Nanyang Technological University, Singapore.

Xie, K.-H., Lu, M.-M., Hu, A.-F. & Chen, G.-H. (2009b) A general theoretical solution for the consolidation of a composite foundation. *Computers and Geotechnics,* 36, 24–30.

Xie, K.-H., Lu, M.M. & Liu, G.B. (2009a) Equal strain consolidation for stone columns reinforced foundation. *International Journal for Numerical and Analytical Methods in Geomechanics,* 33 (15), 1721–1735.

Yee, Y.W. & Raju, V.R. (2007). Ground improvement using vibro replacement (vibro stone columns) – historical development, advancements and case histories in Malaysia. In: *16th Southeast Asian Geotechnical Conference, Kuala Lumpur, Malasia.*

Zayen, V.D.B., Almeida, M.S.S., Marques, M.E.S. & Fujii, J. (2003) Behaviour of the embankment of the sewage treatment plant of Sarapuí (in Portuguese). *Soils and Rocks,* 26 (3), 261–271.

# Annex

Table A.1 Geotechnical characteristics of some soft marine clay deposits in Brazil (Lacerda; Almeida, 1995)

| Soil properties | Plains of Santos (SP) Southeast | Sarapuí (RJ) Southeast | Port of Rio Grande (RS) South | Recife (PE) Northeast | Port of Sergipe (SE) Northeast |
|---|---|---|---|---|---|
| Thickness of clay (m) | <50 | 11 | 40 | 19 | 7 |
| $w_n$ (%) | 90–140 | 100–170 | 45–85 | 40–100 | 40–60 |
| $w_L$ (%) | 40–150 | 60–150 | 40–90 | 50–120 | 50–90 |
| $I_P$ (%) | 15–90 | 30–110 | 20–60 | 15–66 | 20–70 |
| Clay (%) (*) | 20–80 | 20–80 | 34–96 | 40–70 | 65 |
| Specific natural weight (kN/m$^3$) | 13.5–15.5 | 13 | 15–17.8 | 15.1–16.4 | 16 |
| Activity | 1–2.2 | 1.4–2.3 | 0.6–1.0 | 0.4–1.0 | 0.5–1.0 |
| Sensitivity | 4–5 | 4.3 | 2.5 | – | 4–6 |
| Organic matter content (%) | 2–7 | 4–6.5 | – | 3–10 | – |
| $C_c/(1+e_0)$ | 0.33–0.51 | 0.36–0.41 | 0.31–0.38 | 0.45 | 0.31 -0.43 |
| $C_s/C_c$ | 0.09–0.12 | 0.10–0.15 | – | 0.10–0.15 | 0.10 |
| $c_v$ (field)/$c_v$ (lab) | 15–100 | 20–30 | – | – | – |
| $S_u$ (kPa)-Vane | 8–40 | 8–20 | 50–90 | 2–40 | 12–25 |
| $G_{50}/S_u$ | 80 | 87 | – | – | 45–100 |
| $S_u/\sigma'_{vm}$ | 0.28–0.30 | 0.35 | 0.30 | 0.28–0.32 | 0.22 -0.24 |
| $\phi'(°)$ | 19–24 | 25–30 | 23–29 | 25–28 | 26–30 |

Table A.2 Geotechnical characteristics of some soft and very soft clay deposits in Barra da Tijuca and Recreio (RJ).

| Locations | Area 1 | Area 2 | Area 3 | Area 4 | Area 5 | Area 6 | Area 7 | Bedeschi (2004) | Area 8 | SESC/SENAC |
|---|---|---|---|---|---|---|---|---|---|---|
| References | – | Almeida et al. (2008b) | – | Crespo Neto (2004) | Macedo (2004); Sandroni and Deotti (2008) | Baroni (2010) | – | Bedeschi (2004) | Baroni (2010) | Almeida et al. (2001); Crespo Neto (2004) |
| Thickness of the soft to very soft clay(m) | 4–20 | 2–11 | 2–15 | 2–11.5 | 5–16 | 2–21.8 | 1.6–9.5 | 7.5 | 2–8 | 3–12 |
| $w_n$ (%) (1) | 100–488 | 76–913 | 67–207 | 72–410 | 116–600 | 191–670 | 72–1.200 | 102–580 | 56–784 | 72–500 |
| $w_L$ (%) | 148–312 | 86–636 | 40–65 | 23–472 | 100–370 | 147–521 | 88–218 | 97–368 | 67–610 | 70–450 |
| $I_P$ (%) | 80–192 | 59–405 | 20–38 | 11–408 | 120–250 | 95–308 | 47–133 | 42–200 | 47–497 | 47–250 |
| % Clay | 26–54 | 15–60 | 15–51 | – | 32 | 23–93 | 2–36 | – | 14–50 | 28–80 |
| Specific natural weight (kN/m³) | 10.2–13.4 | 10.2–14.0 | 11.9–14.6 | 11–12.4 | 11.6–12.5 | 10.01–12.7 | 10.9–14.9 | 11.2–12.3 | 10.2–16.9 | 12.5 |
| $CR = C_c/(1 + e_o)$ | 0.4–0.8 | 0.22–0.49 | – | 0.27–0.46 | 0.36–0.5 | 0.31–0.54 | 0.11–0.38 | 0.32–0.48 | 0.20–0.63 | 0.29–0.52 |
| $c_v$ (m²/s) × $10^{-8}$ (2) | 0.6–8.8 | 0.3–3.3 | 2.1–49 | 0.1–0.6 | 0.4–1.2 | 0.018–19.8 | 0.6–6.3 | 0.1–19.2 | 0.04–7.5 | 0.17–80 |
| $e_o$ | 3.3–8.2 | 3.0–21.9 | 2.2–4.7 | 3.8–15.0 | 4.8–7.6 | 4.0–12.4 | 1.0–11.6 | 4.3–9.0 | 1.4–10.7 | 2–11.1 |
| $S_u$ (kPa) | 3–38 | 4–18 | 7–41(4) | 3–19 | 5–23 | 2–23 | 2–19 | 1–22 | 4–22 | 7–19 |
| $N_{kt}$ (3) | 4–16 | 4–16 | – | – | 4–9 | 7–17 | – | – | – | – |

(1) Soft clay and peat.

(2) $c_v$ values obtained through piezocone (CPT$_u$) and consolidation test for normally compacted clay. Corrections were carried out for the ch values of piezocone for vertical flow and usually consolidated stretch/portion.

(3) Cone factor = $N_{kt} = (q_t - \sigma_{vo})/S_u$, obtained by correlation vane-piezocone for the deposit, where $q_t$ is the corrected strength measured at the tip of the cone on the piezocone test and $S_u$ is the uncorrected undrained shear strength of the vane test.

(4) Values for the piezocone tests for $N_{kt} = 13$.

Table A.3 Geotechnical characteristics of some soft and very soft clay deposits in the state of Rio de Janeiro (Futai et al., 2008).

| Parameter/Clay | Caju | Santa Cruz (Zona 1) | Santa Cruz (Zona 2) | North shore of Guanabara Bay | Itaipu | Juturnaíba | Uruguaiana | Botafogo |
|---|---|---|---|---|---|---|---|---|
| References | Lira (1988); Lacerda and Cunha (1991) | Aragão (1975) | Aragão (1975) | Aragão (1975) | Carvalho (1980); Sandroni et al. (1984) | Coutinho and Lacerda (1987) | Lins and Vilela (1976) | Lacerda (1980) |
| Thickness of soft clay layer (m) | 12 | 15 | 10 | 8.5 | 10 | 7 | 9 | 6 |
| $w_n$ (%) | 88 | 112 | 130 | 113 | 240 ± 110 | 154 ± 95.6 | 54.8 ± 15.9 | 35 |
| $w_L$ (%) | 107.5 | 59.6 | 125.4 | 122 | 175.4 ± 82.6 | 132.5 ± 43.8 | 71.3 ± 30.0 | 38 |
| $I_P$ (%) | 67.5 | 32 | 89 | 81 | 74.5 ± 30.1 | 63.59 ± 22.1 | 40.5 ± 22.03 | 11 |
| % Clay | – | – | 54 | 35 | – | 60.7 ± 12.74 | 39.4 ± 10.11 | 28 |
| Specific natural weight (kN/m$^3$) | 14.81 | 13.24 | 13.44 | 13.24 | 12 ± 1.85 | 12.5 ± 1.87 | 16.1 ± 1.39 | 17.04 |
| Sensitivity | 3 | 3.39 | 2–6 | – | 4–6 | 5–10 | 3 | – |
| Organic matter content (%) | – | – | – | – | 32.63 ± 20.46 | 19 ± 10.63 | 2.56 ± 1.04 | – |
| $CR = C_c/(1 + e_o)$ | 0.27 | 0.32 | – | 0.26 ± 0.15 | 0.41 ± 0.12 | 0.31 ± 0.12 | 0.31 ± 0.15 | 0.16 |
| $C_r/C_c$ | 0.21 | 0.10 | – | 0.16 ± 0.04 | – | 0.07 ± 0.06 | – | 0.19 |
| $c_v$ (m$^2$/s) × 10$^{-8}$ | | 0.2–18.2 | 0.4 | 0.4 | 5 | 1–10 | | 30 |
| $e_o$ | 2.38 | 3.09 | 3.37 | 2.91 | 6.72 ± 3.1 | 3.74 ± 1.89 | 1.42 ± 0.36 | 1.1 |

*Table A.4* Geotechnical characteristics of some soft and medium clays in the North region

| Parameter/Clay | Low areas of Belém do Pará (Alencar et al., 2001) | | | Porto Cai N'Água – Porto Velho (Marques; Oliveira and Souza, 2008) |
| --- | --- | --- | --- | --- |
| | Very soft organic clay on superficial layer | Variegated clay underlying the 1st layer | Soft to medium dark gray clay below the 1st resistant layer | |
| $N_{SPT}$ | 0–1 | 3–6 | 4–6 | 0–4 |
| $w_n$ (%) | 40–88 | – | – | 31 |
| $w_L$ (%) | 23–58 | 68 | 60 | – |
| $I_P$ (%) | 67.5 | 43 | 27 | 19 |
| % Clay | – | 81 | 61 | < 30 |
| Specific natural weight ($kN/m^3$) | 15–16 | 17.5–18.7 | 17.5–18.5 | 18.9 |
| $C_c$ | 0.8–1.2 | 0.39–0.67 | 0.14–0.285 | – |
| $c_v$ ($m^2/s$) $\times 10^{-8}$ | 5.5–8.5 | – | 3.8–5 | 20–60 |
| $CR = C_c/(1 + e_o)$ | – | – | – | 0.1 |
| $e_o$ | 1.7–2.4 | 0.91–1.19 | 0.89–0.94 | 0.831 |

# Subject index